THE ORGANIC
CHEM LAB
SURVIVAL MANUAL

EIGHTH EDITION

THE ORGANIC CHEM LAB SURVIVAL MANUAL
A Student's Guide to Techniques

JAMES W. ZUBRICK
Hudson Valley Community College

WILEY

John Wiley & Sons, Inc.

For Anne and Zoë,
making the effort worthwhile.

Vice President and Publisher	*Kaye Pace*
Associate Publisher	*Petra Recter*
Editorial Program Assistant	*Catherine Donovan*
Production Services Manager	*Dorothy Sinclair*
Production Editor	*Janet Foxman*
Marketing Manager	*Kristine Ruff*
Creative Director	*Harry Nolan*
Senior Designer	*Carole Anson*
Cover Design	*Wendy Lai*
Illustration Editor	*Anna Melhorn*
Executive Media Editor	*Thomas Kulesa*
Production Services	*Jean Nicolazzo/Aptara®, Inc.*

This book was set in 10/12 Times by Aptara®, Inc. and printed and bound by Courier/Westford. The cover was printed by Courier/Westford.

This book is printed on acid-free paper. ∞

Library of Congress Cataloging-in-Publication Data:
Zubrick, James W.
   The organic chem lab survival manual : a student's guide to techniques / James W. Zubrick. — 8th ed.
   p. cm.
   ISBN 978-0-470-49437-0 (pbk. : alk. paper)
   1.   Chemistry, Organic—Laboratory manuals.   I. Title.
QD261.Z83 2010
547.0078—dc22

2009037069

ISBN 978-0-470-49437-0
Printed in the United States of America

10 9 8 7 6 5 4

PREFACE TO THE EIGHTH EDITION

This eighth edition of the *Survival Manual* again presents the basic techniques of the organic chemistry laboratory, with an emphasis on doing the work correctly the first time. As ever, I have relied on the comments of users and reviewers as a guide to the changes and additions that accompany this eighth edition.

In this new edition, I have made a significant number of additions. I have added to the section on laboratory safety, added a section on green chemistry, and added coverage of planning a synthesis (Chapter 1). After all, if you have to plan a new synthesis, or even modify an existing one, it would be better to incorporate materials and techniques that are less harmful to the environment, create less waste, and are safer to perform in general. Any stoichiometric calculations have also been covered in the sections on notebook keeping (Chapter 2).

The sections on GC and HPLC (Chapters 32 and 33) have been updated in part to reflect data capture and analysis by computer, rather than capture by chart recorder and analysis by less automated means. The section on NMR (Chapter 35) now opens with a presentation of the theoretical basis of the NMR experiment (much as the section on IR does), and includes discussion of some of the consequences of higher-base-frequency instruments, how the FTNMR is developed, and a few new pieces of general information that are especially suitable to the FTNMR experiment.

A section on VIS-UV spectroscopy has been added to the end of the chapter on IR (Chapter 34), using the perhaps flimsy rationale that the instrumentation regimes used in the two techniques have many common elements. A short introduction to some theoretical aspects is followed by solid information both on scanning and CCD instrumentation and on techniques of sample preparation, including the pitfalls of choosing plastic, glass, or quartz cells and how to tell them apart.

Many of the chapters now have exercises. These range from direct, simple questions that help to organize and codify the basic information, to open-ended exercises that can require both a bit of research and a bit of thought, to openly outrageous inquiries designed to drive home a point. While some questions have many possible answers depending on your local laboratory setup, some guidance on finding what the solutions might be can be found at www.wiley.com//college/zubrick. My goal is to reinforce even further the safe and effective implementation of techniques used in the organic chemistry laboratory.

I'd like to thank my reviewers, Scott Allen, University of Tampa; Peter T. Bell, Tarleton State University; Steven M. Bonser, Millersville University; J. R. Dias, University of Missouri–Kansas City; Maged Henary, Georgia State University; Syed

Raziullah Hussaini, University of Louisville; Valerie Keller, University of Chicago; DeeAnne Goodenough-Lashua, University of Notre Dame; Deborah Lieberman, University of Cincinnati; Christopher J. Peeples, The University of Tulsa; Robert Stockland, Bucknell University; and Bruce Toder, University of Rochester, for their comments and suggestions, many of which have been incorporated.

Finally, I'd like to thank Petra Recter, Associate Publisher at John Wiley & Sons, Inc., for her valuable comments and her encouragement in getting out this edition, and Janet Foxman, Senior Production Editor, for seeing that this edition of the *Survival Manual* looks as good as it does. A special thanks to Catherine Donovan, who has helped to pilot this book through a number of editions, and is a wizard in her own right, keeping all of this from flying apart.

J. W. Zubrick
Hudson Valley Community College

SOME NOTES ON STYLE

It is common to find instructors railing against poor usage and complaining that their students cannot as much as write one clear, uncomplicated, communicative English sentence. Rightly so. Yet I am astonished that the same people feel comfortable with the long and awkward passive voice, the pompous "we" and the clumsy "one," and that damnable "the student," to whom exercises are left as proofs. The constructions, which appear in virtually all scientific texts, do *not* produce clear, uncomplicated, communicative English sentences. And students do learn to write, in part, by following example.

I do not go out of my way to boldly split infinitives, nor do I actively seek prepositions to end sentences with. Yet by these constructions alone, I may be viewed by some as aiding the decline in student's ability to communicate.

E. B. White, in the second edition of *The Elements of Style* (Macmillan, New York, 1972, p. 70), writes:

> *Years ago, students were warned not to end a sentence with a preposition; time, of course, has softened that rigid decree. Not only is the preposition acceptable at the end, sometimes it is more effective in that spot than anywhere else. "A claw hammer, not an axe, was the tool he murdered her with." This is preferable to "A claw hammer, not an ax, was the tool with which he murdered her."*
>
> *Some infinitives seem to improve on being split, just as a stick of round stovewood does. "I cannot bring myself to really like the fellow." The sentence is relaxed, the meaning is clear, the violation is harmless and scarcely perceptible. Put the other way, the sentence becomes stiff, needlessly formal. A matter of ear.*

We should all write as poorly as White.

With the aid of William Strunk and E. B. White in *The Elements of Style,* and that of William Zinsser in *On Writing Well* and Rudolph Flesch in *The ABC of Style,* I have tried to follow some principles of technical communication still being ignored in scientific texts: use the first person, put yourself in the reader's place, and—the best for last—use the active voice and a personal subject.

The following product names belong to the respective manufacturers. Registered trademarks are indicated here, as appropriate; in the text, the symbol is omitted.

Büchi®	Büchi Labortechnik, AG, Flawil, Switzerland
Corning®	Corning Glass Works, Corning, New York
Drierite®	W.A. Hammond Drierite Company, Xenia, Ohio

Fisher-Johns®	Fisher Scientific Company, Pittsburgh, Pennsylvania
Kimwipe®	Kimberly-Clark Corporation, Neenah, Wisconsin
Luer-Lok®	Becton, Dickinson and Company, Rutherford, New Jersey
Mel-Temp®	Laboratory Devices, Cambridge, Massachusetts
Millipore®	Millipore Corporation, Bedford, Massachusetts
Swagelok®	Crawford Fitting Company, Solon, Ohio
Teflon®	E.I. DuPont de Nemours & Company, Wilmington, Delaware
Variac®	General Radio Company, Concord, Massachusetts

CONTENTS

CHAPTER 1

SAFETY FIRST, LAST, AND ALWAYS

- *Wear your goggles over your eyes.*
- *If you don't know where a waste product goes—ASK!*
- *Careful reading can prevent failure.*

The organic chemistry laboratory is potentially one of the most dangerous of under-graduate laboratories. That is why you must have a set of safety guidelines. It is a very good idea to pay close attention to these rules, for one very good reason:

The penalties are only too real.

Disobeying safety rules is not at all like flouting many other rules. *You can get seriously hurt.* No appeal. No bargaining for another 12 points so you can get into medical school. Perhaps as a patient, but certainly not as a student. So, go ahead. Ignore these guidelines. But remember—

You have been warned!

1. **Wear your goggles.** Eye injuries are extremely serious but can be mitigated or eliminated if you keep your goggles on *at all times.* And I mean *over your eyes,* not on top of your head or around your neck. There are several types of eye protection available, some of them acceptable, some not, according to local, state, and federal laws. I like the clear plastic goggles that leave an un-broken red line on your face when you remove them. Sure, they fog up a bit, but the protection is superb. Also, think about getting chemicals or chemical fumes trapped under your contact lenses before you wear them to lab. Then don't wear them to lab. Ever.

2. **Touch not thyself.** Not a Biblical injunction, but a bit of advice. You may have just gotten chemicals on your hands in a concentration that is not notice-able, and, sure enough, up go the goggles for an eye wipe with the fingers. Enough said.

3. **There is no "away".** Getting rid of chemicals is a very big problem. You throw them out from here, and they wind up poisoning someone else. Now there are some laws to stop that from happening. The rules were really designed for industrial waste, where there are hundreds of gallons of waste that all has the same composition. In a semester of organic lab, there will be much smaller amounts of different materials. Waste containers could be provided for every-thing, but this is not practical. If you don't see the waste can you need, ask your instructor. When in doubt, *ask.*

4. **Bring a friend.** *You must never work alone.* If you have a serious accident and you are all by yourself, you might not be able to get help before you die. Don't work alone, and don't work at unauthorized times.

5. **Don't fool around.** Chemistry is serious business. Don't be careless or clown around in lab. You can hurt yourself or other people. You don't have to be som-ber about it—just serious.

6. **Drive defensively.** Work in the lab as if someone else were going to have an accident that might affect you. Keep the goggles on because *someone else* is going to point a loaded, boiling test tube at you. *Someone else* is going to spill hot, concentrated acid on your body. Get the idea?

7. *Eating, drinking, or smoking in lab.* Are you kidding? Eat in a chem lab?? Drink in a chem lab??? Smoke, and blow yourself up????

8. *The iceman stayeth, alone.* No food in the ice machine. "It's in a plastic bag, and besides, nobody's spilled their product onto the ice yet." No products cooling in the ice machine, all ready to tip over, either. Use the scoop, and nothing but the scoop, to take ice out of the machine. And don't put the scoop in the machine for storage, either.

9. *Keep it clean.* Work neatly. You don't have to make a fetish out of it, but try to be *neat.* Clean up spills. Turn off burners or water or electrical equipment when you're through with them. Close all chemical containers after you use them. Don't leave a mess for someone else.

10. *Where it's at.* Learn the locations and proper use of the fire extinguishers, fire blankets, safety showers, and eyewash stations.

11. *Making the best-dressed list.* Keep yourself covered from the neck to the toes—no matter what the weather. That might include long-sleeved tops that also cover the midsection. Is that too uncomfortable for you? How about a chemical burn to accompany your belly button, or an oddly shaped scar on your arm in lieu of a tattoo? Pants that come down to the shoes and cover any exposed ankles are probably a good idea as well. No open-toed shoes, sandals, or canvas-covered footwear. No loose-fitting cuffs on the pants or the shirts. Nor are dresses appropriate for lab. Keep the midsection covered. Tie back that long hair. And a small investment in a lab coat can pay off, projecting that extra professional touch. It gives a lot of protection, too. Consider wearing disposable gloves. Clear polyethylene ones are inexpensive, but the smooth plastic is slippery, and there's a tendency for the seams to rip open when you least expect it. Latex examination gloves keep their grip and don't have seams, but they cost more. Gloves are not perfect protectors. Reagents like bromine can get through and cause severe burns. They'll buy you some time, though, and can help mitigate or prevent severe burns. Oh, yes—laboratory aprons: They only cover the *front,* so your exposed legs are still at risk from behind.

12. *Hot under the collar.* Many times you'll be asked or told to heat something. Don't just automatically go for the Bunsen burner. That way lies *fire.* Usually—

No flames!

Try a hot plate, try a heating mantle (see Chapter 17, "Sources of Heat"), but try to stay away from flames. Most of the fires I've had to put out started when some bozo decided to heat some flammable solvent in an open beaker. Sure, there are times when you'll *have* to use a flame, but use it away from all flammables and in a hood (Fig. 1.1), and only with the permission of your instructor.

13. *Work in the hood.* A hood is a specially constructed workplace that has, at the least, a powered vent to suck noxious fumes outside. There's also a

FIGURE 1.1 A typical hood.

safety glass or plastic panel you can pull down as protection from exploding apparatus (Fig. 1.1). If it is at all possible, treat every chemical (even solids) as if toxic or bad-smelling fumes can come from it, and carry out as many of the operations in the organic lab as you can *inside a hood,* unless told otherwise.

14. ***Keep your fingers to yourself.*** Ever practiced "finger chemistry"? You're unprepared so you have a lab book out, and your finger points to the start of a sentence. You move your finger to the end of the first line and do that operation—

   "Add this solution to the beaker containing the ice-water mixture"

And WHOOSH! Clouds of smoke. What happened? The next line reads—

   "very carefully as the reaction is highly exothermic."

But you didn't read that line, or the next, or the next. So you are a danger to yourself and everyone else. Read and take notes on any experiment before you come to the lab (see Chapter 2, "Keeping a Notebook").

15. ***Let your eyes roam.*** Not over to another person's exam paper, but all over the entire label of any reagent bottle. You might have both calcium carbonate and calcium chloride in the laboratory, and if your eyes stop reading after the word "calcium," you have a good chance of picking up and using the wrong reagent. At the very least, your experiment fails quietly. You don't really want to have a more exciting exothermic outcome. Read the entire label and be sure you've got the right stuff.

16. *What you don't know can hurt you.* If you are not sure about an operation, or you have any question about handling anything, *please* ask your instructor before you go on. Get rid of the notion that asking questions will make you look foolish. Following this safety rule may be the most difficult of all. Grow up. Be responsible for yourself and your own education.

17. *Blue Cross or Blue Shield?* Find out how you can get medical help if you need it. Sometimes during a summer session the school infirmary is closed, and you would have to be transported to the nearest hospital.

18. *What's made in Vegas, stays in Vegas.* You're preparing a compound, and you have a question about what to do next. Perhaps your instructor is in the instrument room, or getting materials from the stockroom, or even just at the next bench with another student. Don't carry your intermediate products around; go *a capella* (without accompaniment of beakers, flasks, or separatory funnels filled with substances) to your instructor and ask that she come over and see what you're talking about. Do not ever carry this stuff out of the main lab, or across or down a hallway—ever. A small vial of purified product to be analyzed in the instrument room, sure. But nothing else.

These are a few of the safety guidelines for an organic chemistry laboratory. You may have others particular to your own situation.

ACCIDENTS WILL NOT HAPPEN

That's an attitude you might hold while working in the laboratory. You are *not* going to do anything or get anything done to you that will require medical attention. If you do get cut, and the cut is not serious, wash the area with water. If there's serious bleeding, apply direct pressure with a clean, preferably sterile, dressing. For a minor burn, let cold water run over the burned area. For chemical burns to the eyes or skin, flush the area with lots of water. In every case, get to a physician if at all possible.

If you have an accident, *tell your instructor immediately. Get help!* This is no time to worry about your grade in lab. If you put grades ahead of your personal safety, be sure to see a psychiatrist after the internist finishes.

DISPOSING OF WASTE

Once you do your reaction, since your mother probably doesn't take organic lab with you, you'll have to clean up after yourself. I hesitated to write this section for a very long time because the rules for cleaning up vary greatly according to, but not limited to, federal, state, and local laws, as well as individual practices at individual colleges. There are even differences—legally—if you or your instructor do the cleaning up. And, as always, things do seem to run to money—the more money you have to spend, the more you can throw away. So there's not much point in even trying to be authoritative about waste disposal in this little manual, but there are a few things I have picked up that you should pay attention to. Remember, my classification scheme may not be the same as the one you'll be using. When in doubt, *ask! Don't just throw everything into the sink. Think.*

Note to the picky: The word nonhazardous, as applied here, means relatively benign, as far as organic laboratory chemicals go. After all, even pure water, carelessly handled, can kill you.

How you handle laboratory waste will depend upon what it is. Here are some classifications you might find useful:

1. *Nonhazardous insoluble waste.* Paper, corks, sand, alumina, silica gel, sodium sulfate, magnesium sulfate, and so on can probably go into the ordinary wastebaskets in the lab. Unfortunately, these things can be contaminated with hazardous waste (see the following items), and then they need special handling.

2. *Nonhazardous soluble solid waste.* Some organics, such as benzoic acid, are relatively benign and can be dissolved with a lot of tap water and flushed down the drains. But if the solid is that benign, it might just as well go out with the nonhazardous insoluble solid waste, no? Check with your instructor; watch out for contamination with more hazardous materials.

3. *Nonhazardous soluble liquid waste.* Plain water can go down the drains, as well as water-soluble substances not otherwise covered below. Ethanol can probably be sent down the drains, but butanol? It's not that water soluble, so it probably should go into the general organic waste container. Check with your instructor; watch out for contamination with more hazardous materials.

4. *Nonhazardous insoluble liquid waste.* These are compounds such as 1-butanol (previously discussed), diethyl ether, and most other solvents and compounds not covered otherwise. In short, this is the traditional "organic waste" category.

5. *Generic hazardous waste.* This includes pretty much all else not listed separately. Hydrocarbon solvents (hexane, toluene), amines (aniline, triethylamine), amides, esters, acid chlorides, and on and on. Again, traditional "organic waste." Watch out for incompatibilities, though, before you throw just anything in any waste bucket. If the first substance in the waste bucket was acetyl chloride and the second is diethylamine (both hazardous liquid wastes), the reaction may be quite spectacular. You may have to use separate hazardous waste containers for these special circumstances.

6. *Halogenated organic compounds.* 1-Bromobutane and *tert*-butyl chloride, undergraduate laboratory favorites, should go into their own waste containers as "halogenated hydrocarbons." There's a lot of agreement on this procedure for these simple compounds. But what about your organic unknown, 4-bromobenzoic acid? I'd have you put it and any other organic with a halogen in the "halogenated hydrocarbon" container and not flush it down the drain as a harmless organic acid, as you might do with benzoic acid.

7. ***Strong inorganic acids and bases.*** Neutralize them, dilute them, and flush them down the sink. At least as of this writing.

8. ***Oxidizing and reducing agents.*** Reduce the oxidants and oxidize the reductants before disposal. Be careful! Such reactions can be highly exothermic. Check with your instructor before proceeding.

9. ***Toxic heavy metals.*** Convert to a more benign form, minimize the bulk, and put in a separate container. If you do a chromic acid oxidation, you might reduce the more hazardous C^{6+} to Cr^{3+} in solution and then precipitate the Cr^{3+} as the hydroxide, making lots of expensive-to-dispose-of chromium solution into a tiny amount of solid precipitate. There are some gray areas, though. Solid manganese dioxide waste from a permanganate oxidation should probably be considered a hazardous waste. It can be converted to a soluble Mn^{2+} form, but should Mn^{2+} go down the sewer system? I don't know the effect of Mn^{2+} (if any) on the environment. But do we want it out there?

Mixed Waste

Mixed waste has its own special problems and raises even more questions. Here are some examples:

1. ***Preparation of acetaminophen (Tylenol): a multistep synthesis.*** You've just recrystallized 4-nitroaniline on the way to acetaminophen, and washed and collected the product on your Buchner funnel. So you have about 30–40 mL of this really orange solution of 4-nitroaniline and by-products. The nitroaniline is very highly colored, the by-products probably more so, so there isn't really a lot of solid organic waste in this solution, not more than perhaps 100 milligrams or so. Does this go down the sink, or is it treated as organic waste? Remember, you have to package, label, and transport to a secure disposal facility what amounts to 99.9% perfectly safe water. Check with your instructor.

2. ***Preparation of 1-bromobutane.*** You've just finished the experiment and you're going to clean out your distillation apparatus. There is a residue of 1-bromobutane coating the three-way adapter, the thermometer, the inside of the condenser, and the adapter at the end. Do you wash the equipment in the sink and let this minuscule amount of a halogenated hydrocarbon go down the drain? Or do you rinse everything with a little acetone into yet another beaker and pour that residue into the "halogenated hydrocarbon" bucket, fully aware that most of the liquid is acetone and doesn't need special halide treatment? Check with your instructor.

3. ***The isolation and purification of caffeine.*** You've dried a methylene chloride extract of caffeine and are left with methylene chloride–saturated drying agent. Normally a nonhazardous solid waste, no? Yes. But where do you put this waste while the methylene chloride is on it? Some would have you put

it in a bucket in a hood and let the methylene chloride evaporate into the atmosphere. Then the drying agent is nonhazardous solid waste. But you've merely transferred the problem somewhere else. Why not just put the whole mess in with the "halogenated hydrocarbons"? Usually, halogenated hydrocarbons go to a special incinerator equipped with traps to remove HCl or HBr produced by burning. Drying agents don't burn very well, and the cost of shipping the drying agent part of this waste is very high. What should you do? Again, ask your instructor.

In these cases, as in many other questionable situations, I tend to err on the side of caution and consider that the bulk of the waste has the attributes of its most hazardous component. This is, unfortunately, the most expensive way to look at the matter. In the absence of guidelines,

1. Don't make a lot of waste in the first place.
2. Make it as benign as possible. (Remember, though, that such reactions can be highly exothermic, so proceed with caution.)
3. Reduce the volume as much as possible.

Oh: Try to remember that sink drains can be tied together, and if you pour a sodium sulfide solution down one sink while someone else is diluting an acid in another sink, toxic, gagging, rotten-egg-smelling hydrogen sulfide can back up the drains in your entire lab, and maybe even the whole building.

MATERIAL SAFETY DATA SHEET (MSDS)

The MSDS for any substance is chock-full of information, including but not limited to the manufacturer, composition (for mixtures), PEL (Permissible Exposure Limit), TLV (Threshold Limit Value), boiling point, melting point, vapor pressure, flash point, and on and on and on. These data sheets are very complete, very thorough, and very irrelevant to working in the undergraduate organic chemistry laboratory. Period.

Don't take my word for it. One outfit, Interactive Learning Paradigms Incorporated (www.ilpi.com/msds.index.html), clearly states: "An MSDS reflects the hazards of working with the material in an occupational fashion. For example, an MSDS for paint is not highly pertinent to someone who uses a can of paint once a year, but is extremely important to someone who does this in a confined space 40 hours a week."

And probably less pertinent, if that's even possible, to someone who will work with 1-bromobutane once in a lifetime.

So if you're teaching organic lab, that's one thing. If you're taking organic lab, well, stick to hazard data and references in the other handbooks and you'll be knowledgeable enough.

GREEN CHEMISTRY AND PLANNING AN ORGANIC SYNTHESIS

While it is always good to "reduce, reuse, recycle," unless you're developing new experiments you don't really have any control over these things. But if you have to plan an organic synthesis from the ground up, might as well do it right.

1. *Eschew the older literature!* 'Fraid so. Many places will initially steer you to *Organic Syntheses,* which runs from 1932 to the present, as the syntheses there have been checked and will work as advertised. Unfortunately, for the early work there, and in many other places, being green just wasn't even thought about. So be careful. A historical collection of techniques in a reference with a current copyright date can detail reactions that would not be considered green today.

2. *Teaching over research.* A better place to look is *The Journal of Chemical Education,* rather than the traditional research resources. While a large research group at a large university can have the resources (read *money*) to have toxic materials disposed of properly, "one-man shops" at community colleges are under greater pressure to reduce the costs of waste disposal, and, while they may not be the ones to originally develop a greener method from the high-powered research lab, they certainly exploit it, often in an inspired fashion.

3. *Make what you want, but use what you make.* You'll have to decide on just how much product you'll need to synthesize. And it depends upon the scale of your apparatus.
   • **Microscale.** For a **solid product,** target at least 200 mg. This should be enough for a melting point, an IR, and an NMR, plus some to hand in to show you made it. If you have to, you can easily recover your product from the NMR solvent; IR might be too problematic to bother about. For a **liquid product,** besides the tests, there might be drying and distillation, so about 2 mL might be your target. Don't forget to use the density of the liquid to calculate the mass you'll need to use for your stoichiometric calculations.
   • **Miniscale.** About 5 g for a solid; about 10 mL of a liquid. Just guidelines, now. The consequences of losing product at any stage are greatly reduced. Doesn't mean you should be sloppy with your technique, though.

4. *Plan to lose.* Now that you know how much you're planning to make, assume you won't be making it in a perfect yield. For first-time-this-has-ever-been-done reactions, you might get 40%; if the reaction has been done before, and you have a published procedure with a posted yield, but *you've* never done this before, add a 10% penalty. Then calculate back to get the amount of starting materials you'll need based on this lower yield.

5. *Timing is everything.* Generally, the reaction times shouldn't be reduced. Paradoxically, if you have the time, you can take the time to find out by running the experiment over and over again using different reaction times to find the

best time. If the published procedure uses half-molar quantities (large-scale equipment), and you rework this for microscale, you might reduce the reaction time since the smaller quantity will have lower thermal mass and not need to be heated for as long a time. Maybe.

6. *Use less-toxic materials.* Easy to say; a bit more difficult to do. Some suggestions in no particular order:
    - Do you even need a reaction solvent? Consider direct combination of reagents.
    - Can you replace chlorinated solvents, especially in extraction? You might consider diethyl ether or ethyl acetate.
    - Can you eliminate toxic metals? A shift from a chromium-based to a manganese-based oxidizer in a reaction may help. Organic catalysts can substitute for those based on heavy metals. That sort of thing.

EXERCISES

1. Make a rough sketch of your lab. Mark where the fire extinguishers, fire blanket, eye wash station, and other safety equipment are, as well as where you'll be working.
2. Why shouldn't you work in a laboratory by yourself?
3. Might there be any problems wearing contact lenses in the laboratory?
4. Biology laboratories often have stools. Why might this be foolish in the organic chemistry laboratory?
5. What the heck are the PEL, TLV, and flash point of substances?
6. Google the MSDS for 2-naphthol. Try to select one from Thermo Fisher and another from J. T. Baker/Mallinckrodt. Speculate as to why one says this compound will cause death on inhalation, and the other, well, not so much. Google the MSDS for sugar, also.

KEEPING A NOTEBOOK

- Take notes before lab; make notes during lab.
- Take the notebook to the balance.
- No blank spaces for "future values"; no backfilling.

A **research notebook** is one of the most valuable pieces of equipment you can own. With it you can duplicate your work, find out what happened at your leisure, and even figure out where you blew it. General guidelines for a notebook are as follows:

1. The notebook must be permanently bound. No loose-leaf or even spiral-bound notebooks will do. It should have a sewn binding so that the only way pages can come out is to cut them out. ($8\frac{1}{2} \times 11$ in. is preferred.) **Duplicate carbonless notebooks** are available that let you make removable copies that you can hand in. (Don't forget the **cardboard separator**—or you'll make lots of copies of your latest labwork when your prose writes through to succeeding pages.) And if the pages aren't already numbered, you should do it yourself.

2. *Use waterproof ink! Never pencil!* Pencil will disappear with time, and so will your grade. Cheap ink will wash away and carry your grades down the drain. Never erase! Just draw *one* line through ~~yuor errers~~ your errors so that they can still be seen. And never, never, never cut any pages out of the notebook!

3. Leave a few pages at the front for a table of contents.

4. Your notebook is your friend, your confidant. Tell it:
    a. What you have done. Not what it says to do in the lab book. What you, yourself, have done.
    b. Any and *all* observations: color changes, temperature rises, explosions . . . anything that occurs. Any *reasonable* explanation of *why* whatever happened, happened.

5. Skipping pages is in *extremely* poor taste. It is NOT done!

6. List the IMPORTANT chemicals you'll use during each reaction. You should include USEFUL **physical properties:** the name of the compound, molecular formula, molecular weight, melting point, boiling point, density, and so on. You might have entries for the number of moles and notes on handling precautions. Useful information, remember. The *CRC Handbook of Chemistry and Physics,* originally published by the Chemical Rubber Company and better known as the *CRC Handbook,* is one place to get this stuff (see Chapter 3, "Interpreting a Handbook").

Note the qualifier "useful." If you can't use any of the information given, do without it! You look things up *before* the lab so you can tell what's staring back out of the flask at you during the course of the reaction.

Your laboratory experiments can be classified as either of two major types: a technique experiment or a synthesis experiment. Each type requires different handling.

A TECHNIQUE EXPERIMENT

In a technique experiment, you get to practice a certain operation *before* you have to do it in the course of a synthesis. Distilling a mixture of two liquids to separate them is a typical technique experiment.

Read the following handwritten notebook pages with some care and attention to the *typeset* notes in the margin. A thousand words are worth a picture or so (Figs. 2.1–2.3).

Notebook Notes

1. Use a descriptive title for your experiment. "Distillation." This implies you've done *all* there is in the *entire* field of distillation. You haven't? Perhaps all you've done is "The Separation of a Liquid Mixture by Distillation." Hmmmmmm.

2. Writing that first sentence can be difficult. Try stating the obvious.

3. There are no large blank areas in your notebook. Draw sloping lines through them. Going back to enter observations after the experiment is over is *not professional.* Initial and date pages anytime you write anything in your notebook.

4. Note the appropriate changes in verb tense. Before you do the work, you might use the present or future tense when you write about something that *hasn't happened yet.* During the lab, since you are supposed to write what you've actually done just after you've actually done it, a simple past tense is sufficient.

A SYNTHESIS EXPERIMENT

In a synthesis experiment, the point of the exercise is to prepare a clean sample of the product you want. All of the operations in the lab (e.g., distillation, recrystallization) are just means to this end. The preparation of 1-bromobutane is a classic synthesis and is the basis of the next series of handwritten notebook pages.

Pay careful attention to the typeset notes in the margins, as well as the handwritten material. Just for fun, go back and see how much was written for the distillation experiment, and note how that is handled in this synthesis (Figs. 2.4–2.9).

Once again, if your own instructor wants anything different, do it. The art of notebook keeping has many schools—follow the perspective of your own school.

Notebook Notes

1. Use a descriptive title for your experiment. "n-Butyl Bromide." So what? Did you drink it? Set it on fire? What?! "The Synthesis of 1-Bromobutane from 1-Butanol"—now *that's* a title.

2. Do you see a section for unimportant side reactions? No. Then don't include any.

3. In this experiment, we use a 10% aqueous sodium hydroxide solution as a wash (see Chapter 15, "Extraction and Washing") and anhydrous calcium chloride as a drying agent (see Chapter 10, "Drying Agents"). These are not listed in the Table of Physical Constants. They are neither reactants nor products. Every year, however, somebody lists the physical properties of *solid* sodium hydroxide, calcium

FIGURE 2.1 Notebook entry for a technique experiment (1).

7

01/23/10

Separation of a Liquid Mixture (cont'd)

Obtained liquid unknown #20 from instructor
and dried it over a slight XS of anhydrous
magnesium sulfate. Set up distillation apparatus as
described (p. 6). Started with the smallest flask
to collect fore-run as suggested by instructor.
Filtered unknown into distilling flask with long
stem funnel. Set heat controller to 40.

Instant modification

Mixture finally beginning to boil!

Liquid condensed on thermometer & temperature
reading shot up to 79°C and stabilized at 84°C
in a few seconds. Collected ≈ 2 ml. as fore-run.
Will discard this later. Dropped Thermowell to
remove heat to stop distillation and change receiving
flasks. Started heating again.

Do a bit of work and write a bit of text.

Collected liquid boiling from 81 to 83°C. Changed
receiver as above. When new material came over,
thermometer read 82°C (!) for a few minutes (ml) the
distillation stopped. Temperature began dropping (!)
Turned heat up (60) and mixture started boiling
again — liquid came over @ 123°C. Collected a little
of this and changed receivers again. Added fresh boiling
stone each time heating stopped. Had to label flasks.
So many of them.

01/23/10

FIGURE 2.2 Notebook entry for a technique experiment (2).

8

01/23/10

Separation of a Liquid Mixture (con't).

Flask #	Contents	
1	fore-run	
2	81°C-83°C	fraction
3	82°C-123°C	change-over
4	120°C-123°C	fraction
5	>123°C	residue

Small amount of liquid left-over in the boiling flask can't get over. Dangerous to heat to dryness. Stopped distillation after collecting fraction from 120-123°C (Flask #4)

Cooled distilling flask and poured contents into a 50 ml Erlenmeyer (Flask #5)

Checked cork stoppers for security & have permission to store flasks, properly labelled, in hood until next lab.

Juz   01/23/10

FIGURE 2.3   Notebook entry for a technique experiment (3).

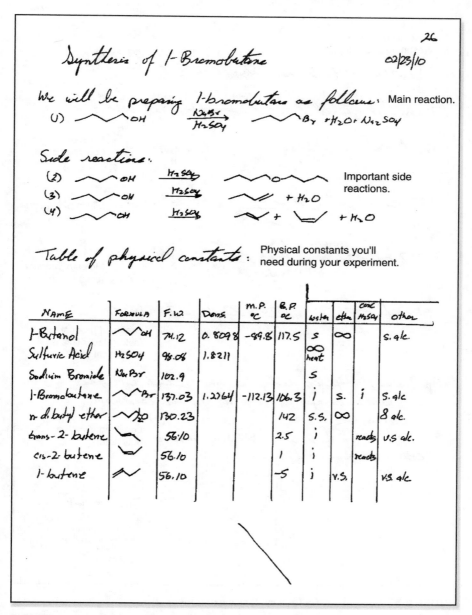

Synthesis of 1-Bromobutane   02/23/10

We will be preparing 1-bromobutane as follows: Main reaction.

(1) ～～～OH $\xrightarrow[H_2SO_4]{NaBr}$ ～～～～Br + H$_2$O + Na$_2$SO$_4$

Side reactions:

(2) ～～～OH $\xrightarrow{H_2SO_4}$ ～～～O～～   Important side
                                                    reactions.
(3) ～～～OH $\xrightarrow{H_2SO_4}$ ～～＝ + H$_2$O

(4) ～～～OH $\xrightarrow{H_2SO_4}$ ～＝ + ＼／ + H$_2$O

Table of physical constants :   Physical constants you'll
                                 need during your experiment.

NAME	FORMULA	F.W.	Dens.	m.p. °C	B.P. °C	water	eth.	conc H$_2$SO$_4$	other
1-Butanol	～～OH	74.12	0.8098	-89.8	117.5	s	∞		s. alc.
Sulfuric Acid	H$_2$SO$_4$	98.08	1.8211			∞ heat			
Sodium Bromide	NaBr	102.9				s			
1-Bromobutane	～～Br	137.03	1.2764	-112.13	106.3	i	s.	i	s. alc
n-dibutyl ether	～～O～～	130.23			142	s.s.	∞		δ alc.
trans-2-butene	＼＝	56.10			2.5	i		reacts	v.s alc.
cis-2-butene	＼＝	56.10			1	i		reacts	
1-butene	＾＝	56.10			-5	i	v.s.		v.s. alc

FIGURE 2.4   Notebook entry for a synthesis experiment (1).

Calculations from
freshman chemistry.

27

Synthesis of 1-Bromobutane (cont'd) 02/23/10

Theoretical yield of ⌇Br based on 17 ml ⌇OH

17 ml ⌇OH × $\dfrac{0.8098 g \ ⌇OH}{1 ml \ ⌇OH}$ = 13.77 g ⌇OH — Mass of liquid from the density.

13.77 g ⌇OH × $\dfrac{1 \ mole \ ⌇OH}{74.12 g \ ⌇OH}$ = 0.1857 mol ⌇OH — Moles of starting material.

0.1857 mol ⌇OH × $\dfrac{1 \ mole \ ⌇Br}{1 \ mole \ ⌇OH}$ = 0.1857 mole ⌇Br — Moles of product (calculated).

0.1857 mol ⌇Br × $\dfrac{137.03 g \ ⌇Br}{1 \ mole \ ⌇Br}$ = 25.44 g ⌇Br — Grams of product (calculated). This is theoretical yield.

Apparatus: Reflux setup as in Zubrick with gas (HBr) trap at top of reflux condenser

air in

OPEN inlet adapter

Distilling adapter — Local modification duly noted.

To Aspirator with trap (slow flow) pulls vapors into water stream

FIGURE 2.5 Notebook entry for a synthesis experiment (2).

28

Synthesis of 1-Bromobutane (cont'd) 02/23/10

Outline of Procedure

1. Place 24.0 g NaBr, 25 ml water and 17 ml 1-butanol in a 250 ml R.B. flask and cool in ice-water bath to < 10°C

2. SLOWLY add 20 ml conc. H_2SO_4 with swirling.

3. Reflux this mixture for 30 minutes.

4. Let mixture cool. Distill mixture - receiver cooled in ice-water bath. Distill mixture until distillate is <u>NOT</u> cloudy.

5. When distillate becomes clear, collect a few drops in small test-tube and add water and shake. If two layers form distill another 5-10 minutes, and repeat this test.
   If two layers do <u>NOT</u> form, distill another 5-10 minutes and quit.

6. Wash distillate with 25 ml H_2O

7. Wash distillate with 15 ml <u>cold</u> conc. H_2SO_4

8. Wash distillate with 15 ml 10% NaOH solution.

9. Dry with anhydrous $MgSO_4$ & gravity filter into small, <u>dry</u> R.B. flask.

10. Distill dried product using <u>dry</u> distillation apparatus.

Juz 02/23/10

FIGURE 2.6 Notebook entry for a synthesis experiment (3).

Synthesis of 1-Bromobutane (cont'd)   29
                                                  02/26/10

Placed 24.0g NaBr, 25ml water and 17ml 1-butanol in a 250 ml R.B. Flask & let cool in an ice-water bath. When solution reached 5°C, added H_2SO_4 with swirling. The mixture warmed up and turned a yellow color.

Set up for reflux with gas trap as on p. 27. Mixture darkened as reflux continued.   Next step performed
                                        while reflux continues.

Placed 15 ml H_2SO_4 in erlenmeyer clamped to cool in ice-water bath for later.   Only a single word recalls the distillation
                                        you had to draw the first time.

After refluxing 30 minutes, let reaction mixture cool to room temp, then put in ice-water bath. There are two distinct layers in the flask, both an orange color. Color may be due to free bromine. One of the layers is my product.

Distilled the mixture, and collected everything that came over up to 100°C. Initially, a cloudy, white liquid came over (H_2O plus organic product?) then clear liquid. Stopped heating distilling flask, removed receiver and replaced it with test-tube in beaker with ice & water. Heated to distill over a few drops of liquid. Added an equal amount of H_2O - shook tube. NO LAYERS FORMED! Replaced receiving flask and distilled for 5 min. more

Poured distillate into a 125 mL separatory funnel & added 25 ml water. Water went into UPPER LAYER; UPPER LAYER is AQUEOUS. Lower layer is organic. SAVE THIS ONE!   Note the recorded observations.

FIGURE 2.7   Notebook entry for a synthesis experiment (4).

Syntheses of 1-Bromobutane (cont'd)   02/26/10   30

Washed product with 15 ml cold conc. H_2SO_4
Solution warmed up as I shook flask!

Washed with 15 ml 10% NaOH. Added 5 ml H_2O
and it stayed in the upper layer. Tested
aqueous layer with red litmus paper and
it turned blue. So organic layer is not
acidic either.

Put product into 50 ml erlenmeyer and added
anhydrous magnesium sulfate in small
amounts with washing. Cloudy product turned
clear and XS unused drying agent was swirling
in the flask. Corked flask and put it away.
02/26/10

03/05/10
    Set up for distillation. Removed thermometer and
thermometer adapter and put long-stem funnel into
the flask. Gravity filtered my product
directly into the distilling flask. Dropped in
boiling stone, replaced thermometer and adapter,
and distilled legend.
Collected all that came over from 100-103°C

03/05/10

FIGURE 2.8   Notebook entry for a synthesis experiment (5).

3)

Synthesis of 1- Bromobutane (cont'd)   03/05/10

~~Weight of labelled vial~~   ~~20.2g~~

Weight of vial and product   36.6 g
Weight of labelled vial   20.2g

Weight of product   16.2g

Put product into clean, weighed, labelled vial.

Product yield: 16.2g

$$\% \ yield = \frac{16.2g}{25.44g} \times 100 = 63.7\%$$

% Yield calculation: your mass of product divided by the calculated mass multiplied by 100.

03/05/10

FIGURE 2.9 Notebook entry for a synthesis experiment (6).

chloride drying agent, and a bunch of other reagents that have nothing to do with the main synthetic reaction. I'm especially puzzled by a listing of solid sodium hydroxide in place of the 10% solution.

4. I'm a firm believer in the use of units, factor-label method, dimensional analysis, whatever you call it. I know I've screwed up if my units are (g 1-butanol)^2/ mole 1-butanol.

5. Remember the huge write-up on the "Separation of a Liquid Mixture by Distillation," drawings of apparatus and all? Well, the line "Distilled the mixture" (Fig. 2.7) is all you need write for a distillation during this synthesis.

CALCULATION OF PERCENT YIELD (NOT YEILD!)

1. *Percent recovery.* When you don't (or can't) know how much you "should get," you're stuck with simply dividing how much you started with into how much you got, and multiplying by 100.

$$\% \text{ Recovery} = \frac{\text{mass of material recovered in grams}}{\text{mass of materials started with in grams}} \times 100$$

If you're isolating a natural product—say, caffeine from coffee—there isn't any way you can possibly know exactly how much caffeine is actually in the sample of coffee you are working on. Yes, you can look up averages and typical amounts and make real good guesses, but they don't necessarily apply to your sample of coffee. Incidentally, the percent recoveries for natural products are exceedingly small; don't panic.

If you're separating and isolating an "unknown mixture," you can't know the masses of the components—that would be telling.

2. *Percent yield.* When you know, or can calculate, how much you "should get," then percent yield is your guy. The actual yield is the easy part—just the weight in grams of your product. The sticking point is usually calculating how much product you'd get if both the reaction and your technique were perfect—the theoretically perfect attainable yield. Since this is just a paper calculation, some call it a calculated yield rather than the more common term, theoretical yield.

$$\% \text{ Yield} = \frac{\text{actual yield of product in grams}}{\text{theoretical yield in grams}} \times 100$$

To calculate the theoretical yield, you have to have a balanced equation, and because chemical reactions happen on a molecular basis, yep, you have to calculate the number of moles of some of the substances you have, determine the limiting reagent (that again!), and use the molecular weights of the compounds involved. Hey, that's why you looked them up and wrote them down in the Table of Physical Constants in your notebook, eh?

Let's look at the reaction and calculations for that synthesis of 1-bromobutane. First and foremost—and this takes a bit of experience (ouch)—the only reagent that is important here in making the product is the 1-butanol. Yes, you need sulfuric acid. Yes, you need sodium bromide. But look at the conversion you're trying to do, and look at the product you're trying to make: 1-butanol to 1-bromobutane. Unless the reaction conditions have been set up very oddly indeed, the limiting reagent will be the important organic compound, here 1-butanol. The sulfuric acid and sodium bromide will be in molar excess (often abbreviated XS). So all of the calculations are based on the quantity of 1-butanol. Of course, the most cautious thing to do is to calculate the number of moles of everything and, taking into account any stoichiometric factors, base your calculations on that.

Because the quantity of the 1-butanol is given as a volume, you have to use the density to convert it to a mass. Then, looking up the molecular weight of 1-butanol in the Table of Physical Constants from your notebook, use that to convert the mass of 1-butanol to the number of moles of 1-butanol. Using the stoichiometric factor—here every 1 mole of 1-butanol gives 1 mole of 1-bromobutane—calculate the number of moles of 1-bromobutane. Finally, using the molecular weight for 1-bromobutane (also in your Table), calculate the theoretical yield, in grams, of 1-bromobutane. It's probably best to use the factor label (or "units") method to do these calculations, and see that the units lead to the correct result as well as the numbers. This is all laid out in Figure 2.5. Note that there isn't even an attempt to calculate the moles of sodium bromide or sulfuric acid, though.

In that example, we begin with 17.0 mL of 1-butanol, and wind up with 16.2 g of the product, 1-bromobutane. To get the percent yield, you have to:

1. *Convert the volume of product into the mass.* You've obtained the 1-butanol as a liquid, measured as a volume. Yes, you can weigh liquids, but we didn't here. Volume-to-mass (and mass-to-volume) conversions use the density. You want to go to mass (g), and you want to get rid of the volume, so

$$17.0 \text{ mL 1-butanol} \times \frac{0.8098 \text{ g 1-butanol}}{1 \text{ mL 1-butanol}} = 13.77 \text{ g 1-butanol}$$

2. *Calculate the moles of 1-butanol.* Here you use the molecular weight. It's given in g/mol, but you have to flip that relationship. If you don't, your units will wind up being g^2/mol instead of mol.

$$13.77 \text{ g 1-butanol} \times \frac{1 \text{ mol 1-butanol}}{74.12 \text{ g 1-butanol}} = 0.1857 \text{ mol 1-butanol}$$

3. *Use the stoichiometric factor to get moles of 1-bromobutane.* Here, 1 mole of 1-butanol is converted to 1 mole of 1-bromobutane. Even though this is pretty simple, keep the units anyway. There will be times when the reaction is not 1:1, and if you don't get the fraction set up correctly—that's all, folks.

$$0.1857 \text{ mol 1-butanol} \times \frac{1 \text{ mol 1-bromobutane}}{1 \text{ mol 1-butanol}} = 0.1857 \text{ mol 1-bromobutane}$$

4. *Calculate the mass of 1-bromobutane.* Now you use the molecular weight of 1-bromobutane, as stated. If you've done everything right, the units work out to be grams of 1-bromobutane.

$$0.1857 \text{ mol 1-bromobutane} \times \frac{137.08 \text{ g 1-bromobutane}}{1 \text{ mol 1-bromobutane}}$$
$$= 25.44 \text{ g 1-bromobutane}$$

5. *Calculate the percent yield.* This is the amount of product you actually obtained, here 16.2 g, divided by the calculated (theoretical) yield of 25.44 g, multiplied by 100.

$$\frac{16.2 \text{ g 1-bromobutane obtained}}{25.44 \text{ g 1-bromobutane calculated}} \times 100 = 63.7\%$$

You usually report this calculated percent yield in whole numbers. Here, 63.7% rounds up to 64%.

6. *Watch out for significant figures.* Chances are you have a number like 63.679245 in your calculator window. You weighed your product to one part in ten (± 0.1), and calculated to one part in one hundred (± 0.01). If the product weight can vary by ± 0.1 g, there's no need for more figures than that.

ESTIMATION IS YOUR FRIEND

You've been asked to weigh 0.07 mole of a reagent called toluene into a 100-mL flask. You punch in the molecular weight and the 0.07, press the key, and get 25.2976496 g. Is this reasonable? Well, the molecular weight of toluene is 92.141 g/mol, and 0.07 is less than 0.1 and 0.1 times about 92 is about 9.2, so the result should have been less than 9.2, not anywhere near 25.

THE ACID TEST

After all of this, you're still not sure what to write in your notebook? Try these simple tests at home.

1. Before lab. "Can I carry out this experiment without any lab manual?"
2. After lab. (I mean *immediately* after; none of this "I'll write my observations down later" garbage.) Ask yourself: "If this were someone else's notebook, could I duplicate the results exactly?"

If you can truthfully answer "yes" to these two questions, you're doing very well indeed.

NOTEBOOK MORTAL SIN

"Bless me, Father, for I have sinned."

"I let someone else borrow my notebook, and he seems to have dropped off the face of the earth. He doesn't come to lab or lecture anymore, and friends can't seem to find him either. My notebook is gone!"

"Well, for your penance . . . I'd say losing your lab notebook is punishment enough."

This doesn't happen often, but it's happened often enough. Lending your lab notebook to anyone else means you really don't have the slightest idea of the purpose of a laboratory notebook. It is not a hand-in project graded for neatness and completeness at the end of a semester. It is the place for you to make notes to yourself about what you are going to do in lab that day and the place you make notes to yourself about what you did and observed in lab that day, too.

What is *another* person going to do with your notebook? Copy your notes made before lab so he—not having even read the lab—could be dangerous not knowing what he's doing? Copy your notes made during the lab so he has the same data and observations you have about an experiment he hasn't done, leeching off your work?

Keep your notebook, and keep it to yourself.

EXERCISES

1. Should what you've written in your notebook in the lab as you are doing the experiment look a lot like what you have written in your notebook as information to help you perform the lab? Why or why not?

2. A schedule of experiments for a laboratory indicates that the 1-bromobutane preparation is paired with the preparation of a tertiary halide, 2-methyl-2-chloropentane, under the single title "Preparation of Alkyl Halides." What do you say to someone who does not have a separate title, introduction, prelab section including a table of physical constants, and such for the preparation of this compound—even though he argues that the experiment schedule doesn't say to read a separate experiment?

3. Drag out your general chemistry book and solve any four stoichiometry problems dealing with a limiting reagent.

4. Unless otherwise specified, solutions are called "weight-weight," even though there may be more than two components, and masses will be used. Is a 10% sodium hydroxide solution made by dissolving 10 grams of the solid in 100 mL of water? Why or why not? What's up with insisting you should dissolve the 10 g of solid hydroxide in only 90 g of water?

- *Read the list of abbreviations to understand the notations.*

You should look up information concerning any organic chemical you'll be working with so that you know what to expect in terms of molecular weight, density, solubility, crystalline form, melting or boiling point, color, and so on. This information is kept in handbooks that should be available in the lab, if not in the library. Reading some of these handbooks is not easy, but once someone tells you what some of the fancy symbols mean, there shouldn't be a problem. Many of the symbols are common to all handbooks and are discussed only once, so read the entire chapter even if your handbook is different. There are at least four fairly popular handbooks, and I've included sample entries for 1-bromobutane and benzoic acid, a liquid and a solid that you might come across in lab, to help explain things.

CRC HANDBOOK

(*CRC Handbook of Chemistry and Physics,* CRC Press, Inc., Boca Raton, Florida, www.crcpress.com.) Commonly called "the *CRC*," as in "Look it up in the *CRC*." A very popular book; a classic. Sometimes you can get the previous year's edition cheaply from the publisher, but it's usually for an order of 10 or more.

Entry: 1-Bromobutane (Fig. 3.1)

1. *No.: 1121.* An internal reference number. Other tables in the handbook will use this number rather than the name.

2. *Name: 1-Bromobutane.* You get a systematic name.

3. *Synonym: Butyl bromide.* Another name for the compound.

4. *Mol. Form: C_4H_9Br.* The empirical formula of 1-bromobutane.

5. *CAS RN: 109-65-9.* The Chemical Abstracts Service registry number. It's a unique identifier for 1-bromobutane.

6. *Mol. Wt.: 137.018.* The molecular weight of 1-bromobutane.

7. *Physical Form: liq.* Yes, 1-bromobutane is a liquid.

8. *mp/°C: –112.6.* The melting point of solid 1-bromobutane. Handbooks report only the *top* of the melting-point range. You, however, should report the *entire* range.

9. *bp/°C: 101.6.* The normal boiling point, at 760 torr, is 101.6°C. Earlier editions also listed 18.8^{30}, which is additional information to tell you that 18.8°C is the boiling point at 30 torr.

10. *den/g cm^{-3}: 1.2758^{20}.* The density of 1-bromobutane. The superscript, here 20, is the temperature (°C) at which the density was taken. These numbers are true densities, and no longer are any of these specific gravities.

11. **n_D: 1.4401^{20}.** This is the index of refraction (see Chapter 29, "Refractometry") obtained using the yellow light from a sodium lamp (the D line). Yes, the tiny 20 means it was taken at 20°C.

12. **Solubility: i H₂O; msc EtOH, eth, ace; sl ctc; s chl.** This is what 1-bromobutane is soluble in. These are listed as abbreviations for the solubility, and then abbreviations for the solvent. Here i H_2O means insoluble in water; msc means miscible—mixes in all proportions—in EtOH (ethanol), eth (ether), and ace (acetone); sl means slightly soluble in ctc (carbon tetrachloride); s means soluble in chl (chloroform). There are many solvents and solubility properties. Here are the abbreviations for some of them:

al	ethyl alcohol	eth	ethyl ether
bz	benzene	chl	chloroform
peth	petroleum ether	w	water
HOAc	acetic acid	MeOH	methyl alcohol
lig	ligroin	ctc	carbon tetrachloride
tol	toluene	ace	acetone
s	soluble	i	insoluble
sl, δ	slightly soluble	msc, ∞	miscible, mixes in all proportions
h	solvent must be hot	v	very

Some solvents have such a long tradition of use that they are our old friends and we use very informal names for them:

alcohol	ethyl alcohol; ethanol
pet. ether	petroleum ether; not a true ether, but a low-boiling (30–60°C) hydrocarbon fraction like gasoline
ether	diethyl ether; ethoxyethane.
ligroin	another hydrocarbon mixture with a higher boiling range (60–90°C) than pet. ether.

Entry: Benzoic Acid (Fig. 3.2)

There are no differences in the entries, with one important exception: Physical Form. Here the abbreviations "mcl lf or nd" refer to the shape of the crystals: monoclinic leaf or needles. Although they are not given here, often colors and recrystallization solvents are listed as well. When you isolate your own solids, you might want to use these descriptors in your own notes. Here are a few:

pl	plates	mcl	monoclinic
nd	needles	rb	rhombus
lf	leaves	ye	yellow
pr	prisms	pa	pale

Thus, pa ye nd (al) means you'll get pale yellow needles when you recrystallize the compound from ethyl alcohol.

Physical Constants of Organic Compounds

No.	Name	Synonym	Mol. Form	CAS RN	Mol. Wt.	Physical Form	mp/°C	bp/°C	den/ g cm^{-3}	n_D	Solubility
1121	1-Bromobutane	Butyl bromide	C_4H_9Br	109-65-9	137.018	liq	−112.6	101.6	1.2758[20]	1.4401[20]	i H_2O; msc EtOH, eth, ace; sl ctc; s chl

FIGURE 3.1 Sample *CRC* entries from the 90th edition. 1-Bromobutane. (Copyright © 2009 from *CRC Handbook of Chemistry and Physics* by D. R. Lide, Ed. Reproduced by permission of Taylor & Francis, a Division of Informa plc.)

Physical Constants of Organic Compounds

No.	Name	Synonym	Mol. Form	CAS RN	Mol. Wt.	Physical Form	mp/°C	bp/°C	den/ g cm^{-3}	n_D	Solubility
712	Benzoic acid	Benzenecarboxylic acid	$C_7H_8O_2$	65-85-0	122.122	mcl lf or nd	122.35	249.2	1.2659[15]	1.504[132]	sl H_2O; vs EtOH, eth; s ace, bz, chl

FIGURE 3.2 Sample *CRC* entries from the 90th edition. Benzoic acid. (Copyright © 2009 from *CRC Handbook of Chemistry and Physics* by D. R. Lide, Ed. Reproduced by permission of Taylor & Francis, a Division of Informa plc.)

LANGE'S

(*Lange's Handbook of Chemistry,* McGraw-Hill Book Company, New York, New York, www.mcgraw-hill.com.) A fairly well-known but not well-used handbook. Some of the entries are similar to those in the *CRC Handbook,* so I'll point out only the interesting differences.

Entry: 1-Bromobutane (Fig. 3.3)

1. *Beilstein Reference: 1, 119.* The Beilstein reference; volume 1, page 119, the original work (not a supplement).
2. *Density: 1.2686_4^{25}.* The 25/4 notation means that this is really a **specific gravity,** not a density, measured at 25°C and referred to as (divided by) the density of water at 4°C.
3. *Flash Point: 18.* The temperature at which the vapor can be ignited by a spark.
4. *Solubility in 100 Parts Solvent: i aq; s alc, bz, eth.* The solubility of one part of 1-bromobutane in 100 parts of solvent. The abbreviations for the solubility and the solvents are given above.

The names listed below the table are synonyms for compounds that would be found if that name were used in the listing. For example, 4-bromobenzyl cyanide should be found right after the entry for α-bromobenzyl cyanide, but the designation b397 means it is listed under that number, which is 4-bromophenylacetonitrile. Finally, there are some chemical structures drawn for compounds whose structure is not easily typeset into a single line.

Entry: Benzoic Acid (Fig. 3.4)

1. *Flash Point: 121 (CC).* Besides the temperature, the notation CC indicates the *known* method by which the flash point was taken: closed cup. There are a few different methods of flash-point determination, and if no method is given, it is assumed to be closed cup.
2. *Solubility in 100 Parts Solvent: 0.29 aq^{25}; 43 alc; 10 bz; 22 chl; 33 eth; 33 acet; 30 CS$_2$.* There are actual gram weights describing the solubility of benzoic acid in various solvents. The first entry means that 0.29 g of benzoic acid will dissolve in 100 g of water at 25°C. Without the superscript, the values are presumed to be at room temperature. The handbook is a bit reticent about letting you know just what room temperature is, but 20°C might be appropriate.

MERCK INDEX

(*The Merck Index,* Merck & Co., Inc., Rahway, New Jersey, www.merck.com.) This handbook is concerned mostly with drugs and their physiological effects. But useful information exists concerning many chemicals. Because of the nature of the listings, I've had to treat the explanations a bit differently from those for the other handbooks.

Physical Constants of Organic Compounds

No.	Name	Formula	Formula Weight	Beilstein Reference	Density, g/mL	Refractive Index	Melting Point, °C	Boiling Point, °C	Flash Point, °C	Solubility in 100 Parts Solvent
b274	1-Bromobutane	$CH_3CH_2CH_2CH_2Br$	137.02	1, 119	1.2686_4^{25}	1.4374^{25}	−112.4	101.6	18	i aq; s alc, bz, eth

FIGURE 3.3 Sample entries from *Lange's Handbook of Chemistry*, 16th edition. 1-Bromobutane. (Copyright © 2005 from The McGraw-Hill Corporation, Inc.)

Physical Constants of Organic Compounds

No.	Name	Formula	Formula Weight	Beilstein Reference	Density, g/mL	Refractive Index	Melting Point, °C	Boiling Point, °C	Flash Point, °C	Solubility in 100 Parts Solvent
b44	Benzoic acid	C_6H_5COOH	122.12	9, 92	1.321		122.4	249	121 (CC)	0.29 aq^{25}, 43 alc; 10 bz; 22 chl; 33 eth; 33 acet; 30 CS_2

FIGURE 3.4 Sample entries from *Lange's Handbook of Chemistry*, 16th edition. Benzoic acid. (Copyright © 2005 from The McGraw-Hill Corporation, Inc.)

Entry: 1-Bromobutane (Fig. 3.5)

1. *Top of page: n-Butyl Chloride 1560.* Just like a dictionary, each page has headings that direct you to the first entry on that page. So, 1560 is not the page number but the compound number for *n*-butyl *n*-butyrate, the last entry on page 255. The actual page number is at the bottom right of the page.

2. *n-Butyl Bromide.* Listed as a substituted butane with the systematic name given as a synonym.

3. *[109-65-9].* The CAS (Chemical Abstracts Service) registry number.

4. *C 35.06%, . . .* Elemental analysis data; the percentage of each element in the compound.

5. *Prepd from . . .* A short note on how 1-bromobutane has been prepared, and references to the original literature (journals).

6. d_4^{25} *1.2686.* The tiny 25 over 4 makes this a specific gravity. Note that the temperatures are given with the *d* and not with the numerical value as in *Lange's* and the *CRC.*

---

n-Butyl Chloride	1560

1553. n-Butyl Bromide. [109-65-9] 1-Bromobutane. C_4-H_9Br; mol wt 137.02. C 35.06%. H 6.62%. Br 58.32%. CH_3-$(CH_2)_3Br$. Prepd from *n*-butyl alc and a hydrobromic-sulfuric acid mixture: Kamm, Marvel, *Org. Synth.* **vol. 1**, 5 (1921); Skau, McCullough, *J. Am. Chem. Soc.* **57**, 2440 (1935).
Colorless liquid. d_4^{25} 1.2686. bp_{760} 101.3, mp –112°, n_D^{25} 1.4398. Insol in water. Sol in alcohol, ether.

Consult the Name Index before using this section.   Page 255

Benzoin	1093

1091. Benzoic Acid. [65-85-0] Benzenecarboxylic acid; phenylformic acid; dracylic acid. $C_7H_6O_2$; mol wt 122.12. C 68.85%, H 4.95%, O 26.20%. Occurs in nature in free and combined forms. Gum benzoin may contain as much as 20%. Most berries contain appreciable amounts (around 0.05%). Excreted mainly as hippuric acid by almost all vertebrates, except fowl. Mfg processes include the air oxidation of toluene, the hydrolysis of benzotrichloride, and the decarboxylation of phthalic anhydride: *Faith, Keyes & Clark's Industrial Chemicals,* F. A. Lowenheim, M. K. Moran, Eds. (Wiley-Interscience, New York, 4th ed., 1975) pp 138-144. Lab prepn from benzyl chloride: A. I. Vogel, *Practical Organic Chemistry* (Longmans, London, 3rd ed, 1959) p 755; from benzaldehyde: Gattermann-Wieland, *Praxis des organischen Chemikers* (de Gruyter, Berlin, 40th ed, 1961) p 193. Prepn of ultra-pure benzoic acid for use as titrimetric and calorimetric' standard: Schwab, Wicher, *J. Res. Natl. Bur. Stand.* **25**, 747 (1940). Review: A. E. Williams in *Kirk-Othmer Encyclopedia of Chemical Technology* **vol. 3** (Wiley-Interscience, New York, 3rd ed., 1978) pp 778-792.

Monoclinic tablets, plates, leaflets. d 1.321 (also reported as 1.266). mp 122.4°. Begins to sublime at ~100°. bp_{760} 249.2°; bp_{400} 227°; bp_{200} 205.8°; bp_{100} 186.2°; bp_{60} 172.8°; bp_{40} 162.6°; bp_{20} 146.7°; bp_{10} 132.1°. Volatile with steam. Flash pt 121 °C. pK (25°)4.19. pH of satd soln at 25°: 2.8. Soly in water (g/l) at 0°

= 1.7; at 10° = 2.1; at 20° = 2.9; at 25° = 3.4; at 30° = 4.2; at 40° = 6.0; at 50° = 9.5; at 60° = 12.0; at 70° = 17.7; at 80° = 27.5; at 90° = 45.5; at 95° = 68.0. Mixtures of excess benzoic acid and water form two liquid phases beginning at 89.7°. The two liquid phases unite at the critical soln temp of 117.2°. Composition of critical mixture: 32.34% benzoic acid, 67.66% water: *see* Ward, Cooper. *J. Phys. Chem.* **34**, 1484 (1930). One gram dissolves in 2.3 ml cold alc, 1.5 ml boiling alc, 4.5 ml chloroform, 3 ml ether, 3 ml acetone, 30 ml carbon tetrachloride, 10 ml benzene, 30 ml carbon disulfide, 23 ml oil of turpentine; also sol in volatile and fixed oils, slightly in petr ether. The soly in water is increased by alkaline substances, such as borax or trisodium phosphate, *see also* Sodium Benzoate.

Barium salt dihydrate. Barium benzoate. $C_{14}H_{10}$-BaO_4-$2H_2O$. Nacreous leaflets. *Poisonous*. Soluble in about 20 parts water; slightly sol in alc.

Calcium salt trihydrate. Calcium benzoate. $C_{14}H_{10}CaO_4$-$3H_2O$. Orthorhombic crystals or powder. d 1.44. Soluble in 25 parts water; very sol in boiling water.

Cerium salt trihydrate. Cerous benzoate. $C_{21}H_{15}$-CeO_6-$3H_2O$. White to reddish-white powder. Sol in hot water or hot alc.

Copper salt dihydrate. Cupric benzoate. $C_{14}H_{10}$-CuO_4-$2H_2O$. Light blue, cryst powder. Slightly soluble in cold water, more in hot water; sol in alc or in dil acids with separation of benzoic acid.

Lead salt dihydrate. Lead benzoate. $C_{14}H_{10}O_4Pb$.-$2H_2O$. Cryst powder. *Poisonous!* Slightly sol in water.

Manganese salt tetrahydrate. Manganese benzoate. C_{14}-$H_{10}MnO_4$-$4H_2O$. Pale-red powder. Sol in water, alc. Also occurs with $3H_2O$.

Nickel salt trihydrate. Nickel benzoate. $C_{14}H_{10}$-NiO_4-$3H_2O$. Light-green odorless powder. Slightly sol in water; sol in ammonia; dec by acids.

Potassium salt trihydrate. Potassium benzoate. C_7H_5 KO_2.$3H_2O$. Crystalline powder. Sol in water, alc.

Silver salt. Silver benzoate. $C_7H_5AgO_2$. Light-sensitive powder. Sol in 385 parts cold water, more sol in hot water; very slightly sol in alc.

Uranium salt. Uranium benzoate; uranyl benzoate. $C_{14}H_{10}$- O_6U. Yellow powder. Slightly sol in water, alc.

Caution: Mild irritant to skin, eyes, mucous membranes.

USE: Preserving foods, fats, fruit juices, alkaloidal solns, etc; manuf benzoates and benzoyl compds; dyes; as a mordant in calico printing; for curing tobacco. As standard in volumetric and calorimetric analysis. Pharmaceutical aid (antifungal).

THERAPCAT (VET): Has been used with salicylic acid as a topical antifungal.

Consult the Name Index before using this section.   Page 181

FIGURE 3.5   Sample entries from *The Merck Index,* 14th edition. (Copyright 2006 Merck and Co., Inc.)

Entry: Benzoic Acid (Fig. 3.5)

1. *Line 2: dracylic acid.* What a synonym! Label your benzoic acid bottles this way and no one will ever "borrow" your benzoic acid again.

2. *Lines 3–6.* Natural sources of benzoic acid; metabolites.

3. *Lines 7–11.* Industrial syntheses of benzoic acid. These are usually not appropriate for your lab bench preparations.

4. *Lines 11–19.* References to the preparation and characteristics of benzoic acid in the original literature (journals).

5. *Structure.* A structural formula of benzoic acid.

6. *Lines 20–36: physical data.* Here are crystalline shape (no longer in the *CRC* or *Lange's*), density (note *two* values reported), sublimation notation, boiling-point data (the subscripts note the pressure in torr), and so on. K at 25° is the ionization constant of the acid; the pH of the saturated solution (2.8 at 25°C) is given. The solubility data (*Soly*) are very complete, including aqueous solutions at various temperatures, a bit about the phase diagram of the compound, and solubility in other solvents. Note that numerical data are given where possible.

7. *Lines 37–62.* Properties of some salts of benzoic acid. Lots of info.

8. *Line 63.* Toxicity data for benzoic acid.

9. *Lines 64–67.* Some commercial uses of benzoic acid.

10. *Lines 68–69.* Therapeutic uses, both human and veterinary, for benzoic acid.

If all of the chemical entries were as extensive as that for benzoic acid, this would be the handbook of choice. Because benzoic acid has wide use in medicine and food production, and it is very important to know the physical properties of drugs and food additives, a lot of information on benzoic acid winds up in the *Index*. 1-Bromobutane has little such use, and the size of the entry reflects this. Unfortunately, many of the compounds with which you come in contact in the organic laboratory are going to be listed with about the same amount of information you'd find for 1-bromobutane, not with the large quantities of data you'd find with benzoic acid.

There's a CD

There's a CD that comes with the 14th edition. Besides sections on organic name reactions and the supplemental tables from the handbook, you get a searchable database that needs the CambridgeSoft Chemfinder Looks pretty sophisticated, but all I did was a few simple searches.

When I entered 1-bromobutane in the search, I got all the data listed in the handbook, but no structure. I guess that's OK, since the handbook doesn't have a structure listed for 1-bromobutane, either. When I entered benzoic acid in the search, I got the entry for acetaminosalol. Entering the CAS registry number for benzoic acid did turn up benzoic acid, though.

THE *ALDRICH CATALOG*

(*Aldrich Catalog Handbook of Fine Chemicals*, Aldrich Chemical Co., Inc., Milwaukee, Wisconsin, www.sigmaaldrich.com.) Not your traditional hard-bound reference handbook, but a handy book nonetheless. The company makes many compounds, some not yet listed in the other handbooks, and often gives structures and physical constants for them. Because Aldrich is in the business of selling chemicals to industry, many industrial references are given.

Although it does seem like an inexpensive way to get a handbook, watch out. Many schools have you run syntheses using cyclopentadiene—either in a Diels-Alder reaction or in the preparation of metallocenes—but there is no entry in the catalog for it. Cyclopentadiene, on standing, dimerizes. So Aldrich can sell you, and publish data for, the dimer, dicyclopentadiene. But if you need information on cyclopentadiene itself, you have to go to a more traditional handbook.

Entry: 1-Bromobutane (Fig. 3.6)

1. *Keywords.* Like in dictionaries, there are keywords at the top of each page. The one on the left-hand page sort of refers to the first entry on that page; the one on the right-hand page sort of refers to the last entry on that page. I say "sort of" because these keywords appear to be brain-dead computer-generated species—no spaces, limited to nine characters, and so on—and are a bit disconcerting to humans expecting real words.

2. *1-Bromobutane.* Here it is listed *strictly* alphabetically—with all the bromo compounds—not as "butane, 1-bromo-," and with only a cross-reference to a butyl bromide.

3. *[109-65-9].* This is the Chemical Abstracts Service (CAS) registry number, as above.

4. *Merck 13, 1552.* The Merck Index, 13th ed., reference; compound number 1552 (see Figure 3.5).

5. *Beil. 1, IV, 258.* Reference to listing in a set of German handbooks called *Beilstein*. Pronounce the German "ei" like the long "I," and stop yourself from saying "Beelsteen" or some such nonsense. 1-Bromobutane is in volume 1 of the 4th Supplement on page 258.

The notation and listings for boiling point, density, melting point, and index of refraction are identical to those of the other handbooks. Because this is a catalog from a chemical supplier, however, there are many more listings for properties involved with the safe handling and storage of chemical compounds. Vapor density (vd), vapor pressure (vp), auto-ignition temperature (ait), lower and upper explosion limit (lel/uel), and flash point (Fp) are listed, as references to the Registry of Toxic Effects of Chemical Substances (RTECS) and Aldrich's own risk statements (R) and safety statements (S). The icons of a flame

1-Bromobutane

Butyl bromide
[109-65-9] CH$_3$(CH$_2$)$_3$Br   FW 137.02   *Merck* **13**,1552; *Beil.* **1**,
IV,258

bp	100-104 °C	vp	40 mm Hg (25 °C)
mp	−112 °C	vp	150 mm Hg (50 °C)
density	1.276 g/mL, 25 °C	ait	509 °F
n$_D^{20}$	1.439	lel/uel(100 °F)	2.8%/6.6%
vd	4.7 (vs air)		

R: 11-36/37/38-51/53   S: 16-26-60   EC No. 203-691-9
Fp: 10 °C (50 °F)   RTECS # EJ6225000

FIGURE 3.6   Bromobutane entry from the Aldrich Catalog 2005–2006.
(Copyright © 2006 Sigma-Aldrich Co.)

(oxidizing), an "X" (irritating), and a dying tree (harmful to the environment) complete the picture.

The safety materials are more easily obtained from the website. Again, Aldrich is interested in selling chemicals, and not helping you write your prelab, so website navigation is a bit more difficult. From the home page, use the search field—typing in the name of the compound—to get to a preliminary listing of the compound in the forms it might be sold. Then select one of these, which may get you a listing of the grades (analytical reagent, technical, etc.) it might be sold in. Finally, click on one of these—usually the analytical reagent—to get to the specifications of the chemical with links to the internal Aldrich safety materials.

Entry: Benzoic Acid (Fig. 3.7)

Because there are no significant differences in the information presented here from that in Figure 3.6, there is just not much to say here.

NOT CLEAR–CLEAR?

One antonym for **clear** is **cloudy.** Another antonym for clear is **colored.** When you say you "obtained a clear liquid," do you mean that it is not cloudy or that it is colorless?

Benzoic acid

[65-85-0] C$_6$H$_5$COOH   FW 122.12   *Merck* **13**,1092; *Beil.* **9**,
IV,273; *Fieser* **1**,49

bp	249 °C	vp	10 mm Hg (132 °C)
mp	121 to 125 °C	ait	1061 °F
vd	4.21 (vs air)		

R: 22-36   S: 26   EC No. 200-618-2   Fp: 121 °C (250 °F)
RTECS # DG0875000

FIGURE 3.7   Benzoic acid entry from the Aldrich Catalog 2005–2006.
(Copyright © 2006 Sigma-Aldrich Co.)

Cloudiness usually means you've gotten water in your organic liquid. Colorless should be self-explanatory. You should always pair the turbidity and color designations:

"a clear, colorless liquid"

"a clear, blue liquid"

"a cloudy, colorless liquid"

"a cloudy, blue liquid"

I use "clear" to mean not cloudy, and "water-white" to mean not colored. Water-white is a designation found in the older chemical literature; **colorless** is more modern.

Is that clear?

INFO ON THE INTERNET

Things have shaken out even more since the last edition. Now I assume you all know your way around the Web, and all that. And finally, I can list a few interesting places without much fear of having them disappear. Even so, I'm only going to make three recommendations: www.google.com, an excellent search engine; www.acs.org, the website of the American Chemical Society; and chemfinder.cambridgesoft.com, where there is an online chemical compounds database. Be careful, though: Besides my having to register after about five searches, entering "c6h5cooh" retrieved "benzoic acid" only second after "salicylaldehyde," along with 113 other compound names, most containing atoms that aren't even in benzoic acid. To its credit, the search engine did retrieve the entry for nitrobenzene when I entered "oil of myrbane." Oh well.

If you have any questions about this Internet stuff or that chemistry stuff, you can e-mail me at

j.zubrick@hvcc.edu

EXERCISES

1. Discuss the possible punishments for someone who has put a pencil mark in a handbook next to a compound you are required to look up. Should the punishment be more or less severe if it is a ballpoint pen mark? Though repulsed by this crass behavior, aren't you secretly glad someone else looked it up and marked the handbook and saved you the trouble of actually looking it up?

2. Look up the physical properties of several of the compounds you will be making in your lab from several different places on the Internet. Do you believe them all to be correct?

3. Look up the Freidel-Crafts acylation reaction of toluene and acetyl chloride. (This is generally Orgo II; stretch yourself!) Look up the product(s) in a handbook. Comment upon the fact that the product is not named as a substituted toluene, even though the acetyl group did substitute for a hydrogen on the toluene ring.

JOINTWARE

■ *Grease the joints to stop sticking.*

Using **standard taper jointware,** you can connect glassware without rubber stoppers, corks, or tubing. Pieces are joined by ground glass connections built into the apparatus (Fig. 4.1). They are manufactured in standard sizes, and you'll probably use ꚍ 19/22.

The symbol ꚍ means **standard taper.** The first number is the size of the joint at the widest point in millimeters. The second number is the length of the joint in millimeters. This is simple enough. Unfortunately, life is not all that simple, except for the mind that thought up this next devious little trick.

STOPPERS WITH ONLY ONE NUMBER

Sounds crazy, no? But with very little imagination, and even less thought, grave problems can arise from confusing the two. Look at Figure 4.2, which shows that all glass stoppers are not alike. Intermixing a ꚍ 19/22 and a ꚍ 19 stopper leads to **leaking joints** through which your **graded** product can escape. Also, the ꚍ 19/22 stopper is much more expensive than the ꚍ 19 stopper, and you may *have to pay money* to get the correct one when you check out at the end of the course. Please note the emphasis in those last two sentences. I appeal to your better nature and common sense. So, take some time to check these things out.

As you can see from Figure 4.2, that single number is the width of the stopper at its top. There is no mention of the length, and you can see that it is too short. The ꚍ 19 stopper *does not* fit the ꚍ 19/22 joint. Only the ꚍ 19/22 stopper can fit the ꚍ 19/22 joint. Single-number stoppers are commonly used with volumetric flasks. Again, they will leak or stick if you put them in a double-number joint.

With these delightful words of warning, we continue the saga of coping with ground-glass jointware. Figure 4.3 shows some of the more familiar pieces of jointware that you may encounter in your travels. They may not be so familiar to you

FIGURE 4.1 Standard taper joints (ꚍ 19/22).

FIGURE 4.2 A ⦙ 19 nonstandard in a ⦙ 19/22 standard taper joint.

now, but give it time. After a semester or so, you'll be good friends, go to reactions together, maybe take in a good synthesis. Real fun stuff!

These pieces of jointware are the more common pieces that I've seen used in the laboratory. You may or may not have *all* the pieces shown in Figure 4.4. Nor will they necessarily be called by exactly the names given here. The point is, *find out* what each piece is, and *make sure* that it is in good condition *before* you sign your life away for it.

ANOTHER EPISODE OF LOVE OF LABORATORY

"And that's $95.48 you owe us for the separatory funnel."

"But it was broken when I got it!"

"Should've reported it then."

"The guy at the next bench said it was only a two-dollar powder funnel and not to worry and the line at the stockroom was long anyway, and . . . and . . . anyway, the stem was only cracked a little . . . and it worked OK all year long . . . Nobody said anything . . ."

"Sorry."

Tales like this are commonplace, and ignorance is no excuse. Don't rely on expert testimony from the person at the next bench. He may be more confused than you are. And equipment that is "slightly cracked" is much like a person who is "slightly dead." There is no in-between. If you are told that you *must* work with damaged equipment because there is no replacement available, you would do well to get it in writing.

FIGURE 4.3 Some jointware.

FIGURE 4.4 More jointware.

HALL OF BLUNDERS AND THINGS NOT QUITE RIGHT

Round-Bottom Flasks

Round-bottom (R.B.) jointware flasks are so round and innocent looking that you would never suspect they can turn on you in an instant.

1. *Star cracks.* A little-talked-about phenomenon that turns an ordinary R.B. flask into a potentially explosive monster. Stress, whether prolonged heating in one spot or indiscriminate trouncing on hard surfaces, can cause a flask to develop a **star crack** (Fig. 4.5) on its backside. Sometimes a crack may be hard to see, but if it is overlooked, the cracked flask may split at the next lab.

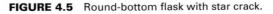

FIGURE 4.5 Round-bottom flask with star crack.

2. *Heating a flask.* Since they are cold-blooded creatures, flasks show more of their unusual behavior when they are being heated. The behavior is usually unpleasant if certain precautions are not taken. In addition to star cracks, various states of disrepair can occur, leaving you with a bench top to clean. Both humane and cruel heat treatment of flasks will be covered in Chapter 17, "Sources of Heat," which is on the SPCG (Society for the Prevention of Cruelty to Glassware) recommended reading list.

Columns and Condensers

A word about **distilling columns** and **condensers:**

Different!

Use the **condenser** as is for **distillation** and **reflux** (see Chapter 19, "Distillation," and Chapter 22, "Reflux and Addition"). You can use the column *with or without column packing* (bits of metal or glass or ceramic or stainless steel sponge—whatever!). That's why the column is wider and has *projections* at the ends (Fig. 4.6). These projections help hold up the column packing if you use any packing at all (see Figure 19.10).

If you jam column packing into the skinny condenser, the packing may never come out again! Using a condenser for a packed column is bad form and can lower your esteem or grade, whichever comes first.

You might use the column as a condenser.
Never use the condenser as a packed column!

The Adapter with Lots of Names

Figure 4.7 shows the one place where joint and nonjoint apparatus meet. There are two parts: a rubber cap with a hole in it and a glass body. Think of the rubber cap as a rubber stopper through which you can insert thermometers, inlet adapters,

Condenser

Projections to support column packing

Wider tube

Distilling column

FIGURE 4.6   Distilling column versus condenser.

REWARD!

Rubber cap with hole

Nonjoint end

Jointware end

THERMOMETER ADAPTER

alias

STRAIGHT ADAPTER

alias

OUTLET ADAPTER

alias

INLET ADAPTER

alias

TUBE ADAPTER

FIGURE 4.7 Thermometer adapter.

drying tubes, and so on. (There's a version that combines this guy with the three-way adapter. See Chapter 5, "Microscale Jointware.")

> *CAUTION! Do not force. You might snap the part you're trying to insert. Handle both pieces through a cloth; lubricate the part (water) and then insert carefully. Keep fingers on each hand no more than 2 in. apart.*

The rubber cap fits over the **nonjoint** end of the glass body. The other end is a **ground-glass joint** and *fits only other glass joints.* The rubber cap should neither crumble in your hands nor need a 10-ton press to bend it. If the cap is shot, get a new one. Let's have none of these corks, rubber stoppers, chewing gum, or any other type of plain vanilla adapter you may have hiding in the drawer.

And remember: Not only thermometers, but **anything** that resembles a glass tube can fit in here! This includes unlikely items such as **drying tubes** (they have an outlet tube) and even a **funnel stem** (you may have to couple the stem to a smaller glass tube if the stem is too fat).

SOCIALLY ACCEPTABLE THINGS TO DO WITH
THE ADAPTER WITH LOTS OF NAMES

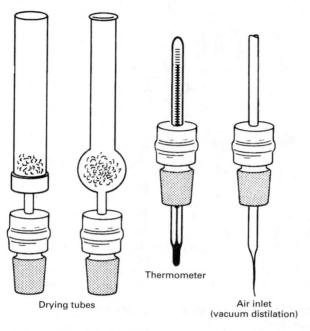

Thermometer

Drying tubes

Air inlet
(vacuum distilation)

FIGURE 4.8   Unusual, yet proper, uses of the adapter with lots of names.

The imaginative arrangements shown in Figure 4.8 are acceptable.

Forgetting the Glass (Fig. 4.9)

The Corning people went to a lot of trouble to turn out a piece of glass that fits perfectly in *both* a glass joint *and* a rubber adapter, *so use it!*

Inserting Adapter Upside Down

This one (Fig. 4.10) is really ingenious. If you're tempted in this direction, go sit in the corner and repeat over and over:

"Only glass joints fit glass joints."

FIGURE 4.9   The glassless glass adapter.

FIGURE 4.10 The adapter stands on its head.

FIGURE 4.11 The adapter on its head without the head.

Inserting Adapter Upside Down *sans* Glass

I don't know whether to relate this problem (Fig. 4.11) to glass-forgetting or upside-downness, since both are involved. Help me out. If I don't see you trying to use an adapter upside down without the glass, I won't have to make such a decision. So, don't do it.

THE O-RING AND CAP BRANCH OUT

Usually, you can find the O-ring, cap, and glass thread hanging out on your favorite pieces of microscale jointware (see Chapter 5, "Microscale Jointware"), but it can show up in some larger setups as well. Figure 4.3 shows a three-way adapter with this kind of inlet at the top, as well as a pear-shaped flask with an inlet at the side. You still have to be careful that you don't drop the rubber ring on the floor where you might not be able to see it very well, that you don't turn the cap so tightly you break the threads, and so on.

GREASING THE JOINTS

In all my time as an instructor, I've never had my students go overboard on greasing the joints, and they never got them stuck. Just lucky, I guess. Some instructors, however, use grease with a passion and raise the roof over it. The entire concept of greasing joints is not as slippery as it may seem.

To Grease or Not to Grease

Generally, you'll grease joints on two occasions: one, when doing vacuum work to make a tight seal that can be undone; the other, when doing reactions with a strong base that can etch the joints. Normally, you don't have to protect the joints during acid or neutral reactions.

Preparation of the Joints

Chances are you've inherited a set of jointware coated with 47 semesters of grease. First wipe off any grease with a towel. Then soak a rag in any hydrocarbon solvent (hexane, ligroin, petroleum ether—and *no flames;* these burn like gasoline) and wipe the joint again. Wash off any remaining grease with a strong soap solution. You may have to repeat the hydrocarbon–soap treatments to get a clean, grease-free joint.

Some suggest using methylene chloride (CH_2Cl_2) to help remove silicone grease. Be very careful with this chlorinated hydrocarbon; dispose of it properly.

Into the Grease Pit (Fig. 4.12)

First, use only enough grease to do the job! Spread it thinly only along the upper part of the joints. Push the joints together with a twisting motion. The joint should

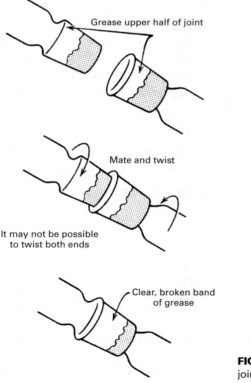

Grease upper half of joint

Mate and twist

It may not be possible to twist both ends

Clear, broken band of grease

FIGURE 4.12 Greasing ground-glass joints.

turn clear from one-third to one-half of the way down the joint. *At no time should the entire joint clear!* This means you have *too much grease* and you must start back at "Preparation of the Joints."

Don't interrupt the clear band around the joint. This is called **uneven greasing** and will cause you headaches later on.

STORING STUFF AND STICKING STOPPERS

At the end of a grueling lab session, you're naturally anxious to leave. The reaction mixture is sitting in the joint flask, all through reacting for the day, waiting in anticipation of the next lab. You put the correct glass stopper in the flask, clean up, and leave.

The next time, the stopper is stuck!

Stuck but good! And you can probably kiss your flask, stopper, product, and grade goodbye!

Frozen!

Some material has gotten into the glass joint seal, dried out, and cemented the flask shut. There are few good cures but several excellent preventive medicines.

Corks!

Yes, corks; old-fashioned, non-stick-in-the-joint corks. If the material you have to store *does not attack cork,* this is the cheapest, cleanest method of closing off a flask.

A well-greased glass stopper *can* be used for materials that attack cork, but *only* if the stopper has a good coating of stopcock grease. Unfortunately, this grease can get into your product.

Do not use rubber stoppers!

Organic liquids can make rubber stoppers swell up like beach balls. The rubber dissolves and ruins your product, and the stopper won't come out either. Ever. The point is:

Dismantle all ground-glass joints before you leave!

CORKING A VESSEL

If winemakers corked their bottles the way some people cork their flasks, there'd be few oenophiles, and we'd probably judge good years for salad dressings rather than wines. You don't just take a new cork and stick it down into the neck of the flask, vial, or what have you. You must press the cork first. Then, as it expands, it makes a very good seal and doesn't pop off.

Before **pressing** or **rolling,** a brand-new cork should fit only about one-quarter of the way into the neck of the flask or vial. Then you roll the lower half of the cork on your *clean* bench top to soften and press the small end. *Now* stopper your container. The cork will slowly expand a bit and make a very tight seal (Fig. 4.13).

Brand-new
unpressed
cork

One-quarter or less
fits in neck

Cork after
pressing
lower half

Half of cork
in neck

FIGURE 4.13 Corking a vessel.

THE CORK PRESS

Rather than rolling the cork on the bench top, you might have the use of a **cork press.** You put the small end of the cork into the curved jaws of the press, and when you push the lever up and down, the grooved wheel rolls and mashes the cork at the same time (Fig. 4.14). Mind your fingers!

Move handle
here to
press
cork

Roll and
press lower
half of
cork

FIGURE 4.14 A wall-mounted cork press.

■ *Don't overtighten the caps.*

MICROSCALE: A FEW WORDS

Microscale is small. How small is it? Well, there's no absolute cutoff, but some suggest 10 g is macroscale, 1 g is semimicroscale, and 0.1 g is truly microscale. Now, 0.1 g of a **solid** is really something. A KBr (potassium bromide) infrared pellet uses 0.01 g (10 mg), and an NMR (nuclear magnetic resonance), which takes five times as much (50 mg), leaves you 40 mg to play with (find melting point, do chemical tests, hand in, and so on). Now, 0.1 g of a **liquid** is really very little. And if you assume that there are about 20 drops/mL for the average liquid and the average liquid has a density of 1 g/mL, you've got two whole microscale drops: one drop for the IR, one for the NMR, and for the boiling point . . . whoops!

Fortunately, most experimenters bend the scale so that you get close to 1 g (approximately 1 mL) when you prepare a liquid product. *That does not give you license to be sloppy!* Just be careful. Don't automatically put your liquid product in the biggest container you have. A *lot* of liquid can lose itself very quickly.

In this chapter I've included drawings of microscale equipment that I've had occasion to use, along with some discussion of the O-ring seals, conical vials, drying tubes, and so on. I've put full descriptions of certain microscale apparatus with the operations in which they're used. So Craig tubes show up with recrystallization; the Hickman still is with distillation.

UH-OH RINGS

Take a quick look at the floor of your organic chemistry laboratory. Now imagine yourself dropping a tiny, rubbery, brown or black O-ring on it. Uh-oh! The floors where I work are a common type of brown-, black-, and white-spotted terrazzo, and it's tough to find dropped O-rings. Make sure your hands are way over the bench top when you're monkeying with these things.

THE O-RING CAP SEAL

In the old days, you would put two glass joints together and, if you wanted a vacuum-tight seal, you'd use a little grease. These days, you're told to use a plastic cap and an O-ring to get a greaseless, high-tech seal. Maybe. Not all O-ring cap seal joints are created equal.

Skinny Apparatus

If you can drop the O-ring over the apparatus from the top, and it'll slide down to the bottom, such as fitting an air condenser to a conical vial, it's very easy to set up:

1. Get a conical vial (appropriate size) that'll fit on the bottom end (male joint) of the wide-bottom air condenser (Fig. 5.1).
2. Drop the O-ring over the top of this air condenser.

2. Drop on cap

1. Drop on O-ring

3. Screw the cap down

FIGURE 5.1   O-ring cap seal on skinny apparatus.

3. Drop the cap over the top of this air condenser.

4. Carefully screw the cap down onto the threads of the vial.

Not-So-Skinny Apparatus

Unfortunately, you *can't* drop an O-ring and cap over a water-jacketed condenser, Hickman still, and so on. What to do? Using the water-jacketed condenser as an example, go from the bottom up:

1. Get a conical vial (appropriate size) that'll fit on the end (male joint) of the condenser (Fig. 5.2).

2. Take the vial *off* the end.

3. Put a plastic cap onto the male joint and hold it there.

4. Push an O-ring up over the male joint onto the clear glass. The O-ring *should* hold the cap up. (Fingers may be necessary.)

5. Put the conical vial *back* on the male joint.

6. Carefully screw the cap down onto the threads of the vial.

Sizing Up the Situation

There have been several variations of the O-ring cap seal and you might still come across some of them:

1. Hold up plastic cap and O-ring

2. Drop this onto vial

3. Screw the cap down

FIGURE 5.2 O-ring cap seal on husky apparatus.

1. *A 3- to 5-mL conical vial and ⊤ 14/10 joints.* The plastic caps for these sizes have holes cut in them that fit *extremely* tightly. They make a really good seal, and I suspect you would break the glass before *ever* pulling one of these babies apart. Once, when dismantling one of these joints, I caught the O-ring in the cap, and as I pulled the cap off the joint it sliced the O-ring in two. Watch it.

2. *A 0.3- to 1.0-mL conical vial and ⊤ 7/10 joint* (Fig. 5.3). The ⊤ joint rattles around in the hole in the plastic caps for this size. As a consequence, whenever you tighten these joints, the O-ring squeezes out from under the cap. Yes, the joint's tight, but gas-tight? You can pull the joints apart with modest effort, and, if the O-ring has squeezed out entirely, forget the cap as a stabilizing force.

3. *A 0.1-mL conical vial and ⊤ 5/5 joint.* For these sizes, again, the hole in the cap is snug around the male joint so that when you screw down the cap, the O-ring doesn't squeeze out. This joint also appears to be very tight.

← O-ring squeezes out

0.3- to 1.0-mL vial and ⊤ 7/10 joint hidden by plastic screw cap

FIGURE 5.3 O-ring squeezes out with ⊤ 7/10 joint.

Why I Don't Really Know How Vacuum-Tight These Seals Are

I'll break here and let Kenneth L. Williamson of *Macroscale and Microscale Organic Experiments* (D. C. Heath & Co., Lexington, MA, 1989, p. 102) take over. Here he is on the subject of microscale vacuum distillation assemblies: "On a truly micro scale (<10 mg) simple distillation is not practical because of mechanical losses."

That's clear enough. You lose lots of your microscale product with a reduced-pressure (vacuum) distillation. As far as I know, no microscale laboratory manual has anyone perform this. Yes, you might be asked to remove solvent under reduced pressure, but that's neither a reduced-pressure distillation nor uniquely microscale.

So, since nobody actually does any microscale reduced-pressure distillations, and since with certain sizes of equipment the O-rings can squeeze out from under the plastic caps, and since I don't use these seals for reduced-pressure (vacuum) distillation anyway (I switch to a semimicro apparatus with real ℑ 14/20 joints and—gasp!—grease), I don't know how vacuum-tight these seals really are.

THE COMICAL VIAL (THAT'S *CONICAL!*)

The **conical vial** (Fig. 5.4) is the round-bottom flask of the microscale set with considerably more hardware. When your glassware kit is brand-new, every vial has a plastic cap, O-ring, and plastic disk to match. After one semester, every vial *needs* this stuff. When you check in, make sure that every conical vial has a cap that fits and that you have at least one O-ring for each and a plastic disk that just fits inside this cap.

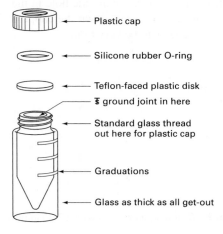

Plastic cap

Silicone rubber O-ring

Teflon-faced plastic disk

ℑ ground joint in here

Standard glass thread out here for plastic cap

Graduations

Glass as thick as all get-out

FIGURE 5.4 The conical vial.

The Conical Vial as Vial

You would think that with an O-ring and a Teflon-faced plastic disk wedged under the cap, this would make a very tight seal. Think again. One of my students made *n*-hexylamine as a special project and packaged it in one of these vials; some of the amine leaked out, air got in, and the amine oxidized to a lovely brown color. You can still use them if you want, but be careful:

1. Check to see that the Teflon-faced plastic disk doesn't have any holes in it. Sometimes you pierce a disk with a syringe needle to add something to a setup. You use the disk as a puncturable seal. With a puncture or two, the seal won't be as good as with a fresh disk. Just to be sure, get a new one if your disk has had lots of punctures.

2. The Teflon side to the chemicals, please. Especially if the compound might attack the plastic disk. Teflon is *inert;* the plastic is *ert.*

3. The O-ring goes close to the cap. Otherwise, it can't press the disk down to the glass.

4. Now the cap. Screw this baby down and you should, *should,* mind you, have a very well-sealed vial.

Laboratory gorillas please note:
Overtightening caps breaks the threads.
Thank you.

Even so, watch for leaks! Sometimes you're asked to perform an extraction in one of these vials, and you decide to shake the bejesus out of the vial to mix the two layers (see Chapter 16, "Extraction and Washing: Microscale"). After you clean your fingers, you *will* be more gentle when you repeat the experiment, won't you?

Packaging Oops

Did you get the weight of the product so that you could calculate the yield? ("What? *Calculation* in organic lab?") Fat lot of good that tight seal will do for you now that you have to undo it to weigh your product. And you'll lose a lot of your product on the walls of the vial. Nothing but problems, especially in microscale.

Tare to the Analytical Balance

You can avoid losing your product and save a bit of time if you weigh the empty vial *before* you put your product in it. Microscale quantities, unfortunately, may require a high-precision **analytical balance** rather than a **top-loading balance,** and you should be aware of at least one thing:

Analytical balances are very delicate creatures.
Only closed containers in the balance box!

Now, I don't know if you have a hanging pan analytical balance with dial-up weights or a fancy electronic model, and I don't care. Just keep your noxious organic products in closed containers, OK?

The Electronic Analytical Balance

1. Open your notebook and, on one line, write "Mass of empty container," and get ready to write in the mass. What? You didn't bring your notebook with you to record the weights? You were going to write the numbers down on an old paper towel? Go back and get your notebook.

2. Turn the balance on. If it's already on, zero the balance. See your instructor for the details.

3. Open the balance case door, and put the empty vial with cap and O-ring and plastic disk on the balance pan. Close the balance case door.

4. Wait for the number to stop jumping around. Even so, electronic digital balances can have a "bobble" of one digit in the least significant place. Write down the most stable number.

5. Take the vial out of the balance case. *Now* fill the vial with your product. Close it up. No open containers in the balance case.

6. Write this new number down. Subtract the numbers. You have the weight of your product and the minimum of loss.

"What if I have a **tare bar** or **tare knob**? Why not use this to zero the balance before putting in product? That way, I don't have to subtract—just get the weight of the product directly." Sure. Now take some of your product out of the vial to use in, say, another reaction. How much did you take? What? You say you already put it into the other reaction vial, and it's all set up and you can't take product out to weigh it? What? You say you think you'll lose too much product reweighing it in yet another vial before you commit it to your reaction? Did you record the tare of the product vial? No? One subtraction too many for you, huh?

But if you did record the weight of the empty vial, cap, and O-ring, all you need do is remove some product, close the vial up, weigh it again, and subtract *this* weight from the sum of the tare and original product mass.

"What if I have a top-loading balance without a cage—must I have only closed containers on it while weighing?" Yes. Cuts down on the mess. An open container on a balance pan is usually an irresistible invitation to load product into it while on the pan, spilling the product onto the pan.

"What about weighing paper?" What *about* weighing paper? You *always* fill or empty your containers *away from the balance pan,* and you always weigh your samples *in these closed containers,* so what about weighing papers?

Heating These Vials

Sources of heat vary according to what you have available. Some use a can-type heating mantle, only it's filled with sand. You raise or lower the vial in the sand pile

FIGURE 5.5   Hot bath for conical vial; just happens to be a reflux.

to set the temperature you want. Others use a **crystallizing dish** on top of a **hot plate with magnetic stirrer** (Fig. 5.5). You put the vial flat on the crystallization dish and add sand. You can monitor the temperature of the vial roughly by sticking a thermometer in the sand.

THE MICROSCALE DRYING TUBE

The **microscale drying tube** (Fig. 5.6) is just a right-angled glass tube with a ⊤ joint on one end. You put a small cotton plug in first and then load in the drying agent. Finish up with another cotton plug.

If you let these sit around too long, the drying agent cakes up, and you could break the tube hacking the stuff out. It's actually cheaper to empty the tube after

FIGURE 5.6   A microscale drying tube.

using it. If you can't hack that, soak the tube overnight (or so), and dissolve out the caked drying agent.

GAS COLLECTION APPARATUS

Several microscale kits have **capillary gas delivery tubes.** With this tube attached, you can collect gaseous products.

1. Put a cap on your **gas collection reservoir.** You can calibrate the tube by adding 2, 3, or 4 mL of water and marking the tube. Some have calibration lines. Check it out (Fig. 5.7).

2. Get a 150-mL beaker and fill it with water to about 130 mL. A 100-mL beaker filled to 80 mL will work, but it'll be a bit tight.

3. Now fill the **gas collection reservoir** with water, put a finger over the open end, invert the tube, and stick the tube (and your fingers) under the water. Remove your finger. See that the gas collection reservoir remains filled.

4. Angle the straight end of the **capillary gas delivery tube** under and into the gas collection reservoir (Fig. 5.8). Believe it or not, the gas collection reservoir, sitting atop the capillary gas delivery tube, is fairly stable, so you can put this aside until you need it.

5. When you're ready, use a ring with a screen to support the beaker at the right height to connect the joint end of the capillary gas delivery tube to whatever you need to. Direct connection to a conical vial usually uses an O-ring cap seal; connection to a condenser may not (Fig. 5.9). Note that I've drawn the water connections to the condenser at an angle. Although the drawing is two-dimensional, your lab bench is not. You may have to experiment to find an

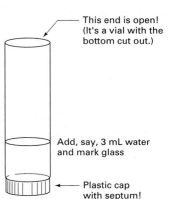

This end is open!
(It's a vial with the bottom cut out.)

Add, say, 3 mL water and mark glass

Plastic cap with septum!

FIGURE 5.7 Calibrating the gas collection reservoir.

FIGURE 5.8 Setting up for gas collection.

arrangement that will accommodate the ring stand, hot plate, and other paraphernalia. Don't get locked into linear thinking.

Generating the Gas

Usually, you warm the conical vial somehow to have the reaction generate the gas you want to collect.

Do not stop heating once you start!
Water will get sucked back into the reaction mixture
as the apparatus cools!

Watch the flow of gas into the collection bottle. If you increase the heat and no more gas comes out, the reaction is probably done.

Remove the capillary gas delivery tube first.
Then remove the heat.

Of course, some gaseous reaction product, a few microliters perhaps, will still come off the mixture. Don't breathe this. And you *did* carry this out in a hood, no?

If your capillary gas delivery tube was connected to the top of a condenser (Fig. 5.9b), just pick up the beaker, collection vial, and capillary gas delivery tube, as a unit, off the condenser, and set it down. Then attend to the reaction.

If your capillary gas delivery tube was connected directly to a conical vial with an O-ring cap seal (Fig. 5.9a), you have to undo the seal on a hot vial (!) to lift the entire system off as a unit. And you can't let the vial cool with the capillary gas delivery tube under water, or water will be sucked back into the vial. The best I've been able to do is carefully lift the gas collection vial off the gas delivery tube (watch the water level in the beaker: If it's too low, the water in the collection vial will run out!) and then lower the beaker with the collection vial in it away from the capillary gas delivery tube. There's got to be a better way, but I don't know it yet. Ask your instructor.

FIGURE 5.9 (*a*) Gas collection directly from a conical vial. (*b*) Gas collection from a condenser.

Isolating the Product

With the **gas collection reservoir,** you have an O-ring cap seal with a Teflon-faced plastic insert. Pierce the insert with a hypodermic syringe and draw off some of the product. Usually, you put this gas through gas chromatography (see Chapter 30, "Gas Chromatography," for particulars) for analysis.

CHAPTER

- *Your thermometer should work.*
- *No glassware even chipped.*

An early edition of this book illustrated some equipment specific to the State University of New York at Buffalo, since that's where I wrote it. Now it's a few years later, and I realize that you can't make a comprehensive list. SUNY Buffalo has an unusual pear-shaped distilling flask that I've not seen elsewhere. The University of Connecticut equipment list contains a Bobbitt filter clip that few other schools have picked up. So if you are disappointed that I don't have a list and drawing of every single piece of equipment in your drawer, I apologize. Only the most common organic lab equipment is covered here. Ask your instructor, "Whatizzit?" if you do not know. I used to assume (check previous editions if you doubt me) that you remembered Erlenmeyer flasks and beakers and such from freshman lab. But it's been *suggested* that I put drawings of even these beasties in this section anyway. OK by me. But I'll still be discussing the other apparatus as it comes up in the various techniques. This might force you to read this book before you start in the lab. So, check out the new and improved (yeah, right) Figure 6.1. Not all the mysterious doodads in your laboratory drawer are shown, but lots of old friends from your freshman chem lab are there, too. Just in case you forgot.

FUNNELS, AND BEAKERS, AND FLASKS—OH MY!

While you are checking in, there are things about these pieces of nonjointware apparatus you should check out. Make sure that the pouring spouts on your beakers aren't cracked. This can happen when you nest the beakers—putting smaller ones inside larger ones—and the spout of the smaller beaker bonks up against the inside wall of the larger beaker and breaks.

The easiest, most paradoxical way to tell if you have broken beaker spouts that can cut you is to slowly and carefully—that's slowly and carefully—run your finger or thumb over the spout. Any roughness and it's damaged. Get a new one.

Slowly and carefully do the same with the rest of the beaker lip. Any damage—get a new one. Check the lip of flasks, spout and lip of graduated cylinders, and rim and angled opening at the stem of your glass funnels.

Check the bottom of the beakers and flasks where they bend, too. Cracks often show up here where the glass is stressed the most.

THE FLEXIBLE DOUBLE-ENDED STAINLESS STEEL SPATULA

This is not about the scoopula. Scoopulas are fine for freshman chemists, but real "organikers" would be lost at their benches without the flexible double-ended stainless steel spatula.

Erlenmeyer flask   Beakers   Filter flask   Graduated cylinders

Long-stem funnel   Powder funnel   Buchner funnel   Hirsch funnel

Drying tubes   Scoopula   Steam bath   A real spatula

FIGURE 6.1 Some stuff from your lab drawer.

Do you have a solid product on the walls of an Erlenmeyer flask? Scrape it off the glass and onto the straight edge of one of the spatula paddles. Solid product stuck on the walls of a round-bottom flask? With one of the paddles bent—sort of curled, actually—you can easily scrape the solid down into a pile at the bottom of the flask, then lift it out with this kind of spatula. Try either of these common operations with a scoopula, and you will fail (Fig. 6.2).

Scrape product from straight
side of Erlenmeyer flask

Scrape product from round
side of R.B. flask with
personally curved end

FIGURE 6.2   Flexible double-ended
stainless steel spatula at work.

I've seen two variants of the flexible double-ended stainless steel spatula. The first version is made of really heavy steel and the paddles can't bend. I believe the word *pointless* is preferred to *useless* here.

The second version has only one long paddle. The other end is a shorter paddle that comes down to a point. That pointed end is excellent for sticking the big rubber band on your lab goggles back under the molded slots when the band pulls out.

PIPET TIPS

CHAPTER

- *Solvents can blow out of a pipet and rubber bulb. Careful.*

- *Pipets filter small amounts of liquid product that might disappear on a gravity filter.*

The Pasteur pipet (Fig. 7.1) is really handy for *all* scales of laboratory work, not just microscale, but it gets the most use in microscale. It usually comes in two sizes, a 9-in. (approximately 3-mL) pipet and a $5\frac{3}{4}$-in. (approximately 2-mL) pipet.

Although they look like eyedroppers, they aren't. The drop size of a Pasteur pipet is very different from that of a traditional eyedropper. I took the average mass of 20 drops of room temperature cyclohexanone and found that an eyedropper delivered about 0.0201 g and a Pasteur pipet delivered about 0.0137 g. The conventional wisdom in these things says there are 20 drops/mL, but these are eyedropper drops, not pipet drops.

PRE-PREPARING PASTEUR PIPETS

Figure 7.1 shows a rough calibration chart for both sizes.

9" pipet

$5\frac{3}{4}$" pipet

FIGURE 7.1 Pasteur pipets in all their glory.

Calibration

You should get an idea about how high in the pipet 0.5, 1.0, 1.5, and 2.0 mL of liquid are. Just draw measured amounts into the tube and look. I'd mark *one* pipet with all these volumes and keep it as a reference. Marking your working pipets can get messy, especially when the markings dissolve into your product. And scratching a working pipet with a file is asking for trouble. Get the scratch wet, a little lateral pressure, and—snap—that's the end of the pipet.

Operation

I know I'm not supposed to have experiments in this book, but here's something I strongly suggest you carry out (translation: Do it!). Work in a hood. Wear goggles. Don't point the end of a pipet at anyone, yourself included.

Experiment: Put a few milliliters of diethyl ether in a test tube. Fit a Pasteur pipet with a rubber bulb. Warm the pipet in your hand for about a minute. Now draw some ether up into the pipet and watch.

If this went at all well, the ether probably came squirting out the end of the pipet. The ether evaporated in the pipet; then vapor filled the bulb and pushed the liquid out. Lots of liquids can do this, and if you've never pipetted anything but water before, the behavior of these liquids is quite striking.

Amelioration

There are two "fixes" to the vapor pressure problem:

1. Prewet the pipet with the solvent you'll have in it. Pull in and then expel the solvent two or three times. Once this vapor fills the pipet, the loss of solvent (as previously noted) is much slower, if not stopped. Even so, *be careful*. Don't squeeze or release the bulb rapidly; don't let your hand preheat or heat the pipet while handling it.

2. Stick a cork in it. A cotton plug, really. Mayo *et al.* suggest that you push a small cotton ball way down into the narrow part of the pipet with a copper wire, rinse the cotton with 1 mL of methanol and then 1 mL of hexane, and let the cotton dry before use (Fig. 7.2a). On *all* your pipets. Well, sure, once you get good at it, it won't take the whole lab period to prepare them. Of course, you could just get good at being careful with this solvent pressure phenomenon.

I'm told a second advantage of putting cotton in all the pipets is that every time a transfer is made, the material is automatically filtered. Pull material into the pipet, and filth gets trapped on the cotton plug. Squirt the material out, and *all* the filth stays on the cotton. None *ever* washes back out. Yep. Sure. I believe, I believe. . . . In any case, don't use too much cotton! A little here goes a long way.

FIGURE 7.2 (*a*) Pasteur filter pipet—Mayo. (*b*) Pasteur filter pipet—Zubrick.

PIPET CUTTING

From time to time you might need to shorten the long tip of the Pasteur pipet so that the narrow portion is, say, 2–3 cm rather than the 8 cm or so of the 9-in. pipet.

1. Get a sharp triangular file or, better yet, a sharp glass scorer.
2. Support the narrow part of the tip on a raised solid surface such as a ring-stand plate.
3. Carefully draw the cutter back toward you while you rotate the pipet away from you. Don't press down too much; you'll crush the tip.
4. Wet the scratch.
5. Using a towel to hold the pipet and to keep from cutting yourself (although the small diameter of the tip makes disaster a bit less likely), position your thumbs on either side of the scratch, with the scratch facing away from you.
6. Read the next two steps before doing anything.
7. Exert force to pull the pipet apart. That's right. Pull first. (The pipet might part here.)
8. If it's still together, while you're *still pulling,* push both thumbs outward.
9. If it hasn't parted yet, either get help from your instructor or don't fool with this one anymore. Get another pipet and try again.
10. Carefully fire-polish the cut end.

PIPET FILTERING—LIQUIDS

Real pipet filtering will trap all of the solid materials. For this you need two pipets.

1. Get a small piece of cotton and roll it into a ball that's just about the size of the large end of the pipet. Now, using the narrow end of *another* pipet, push the cotton ball just past the constriction, down into the first pipet. Then use a wire to push the cotton ball down into the pipet. Give it three to five jabs—much as if you were killing the nerves in a tooth during a root canal (Fig. 7.2*b*).

   If it's at all possible, unravel, unbend, and straighten out a wire coat hanger and use it to slowly push the cotton ball down. The thick coat hanger wire is more effective at pushing a cotton ball down the pipet but doesn't have the imagery of root canal work.

   You don't have to push the cotton all the way into the narrow part of the tip. Just wedge it firmly in the narrows.

   Trial and error have shown that an ordinary coat hanger wire is small enough to go down the pipet; large enough not to just push through the cotton ball, leaving the ball suspended somewhere in the middle of the pipet; and stiff enough to do the job admirably. The larger-diameter coat hanger wire will not, repeat not, be able to force the cotton way down into the tip, but you don't really want to do that anyway.

2. Get a vial or other receptacle ready to receive your filtered product.

3. Fill the filter pipet from another pipet.

4. Let the liquid drain out of the filter pipet into your clean vial. Note that you could—could, mind you—put a rubber bulb onto the filter pipet and use the bulb to help push the solution out. Watch out:

   a. If you push the solution through the cotton too rapidly, you could force filth through the cotton and into the cleaned product.

   b. Don't *ever* let up on the rubber bulb once you start. You'll suck air and possibly filth off the cotton back into the pipet. In a few cases, where people didn't jam the cotton into the tip, but had the cotton as a large ball wedged in the narrowing part, the cotton would be sucked back into the pipet, and the entire solution, filth and all, would run right out of the pipet.

   c. You could—this is rare—blow the cotton jammed in the tip right out.

PIPET FILTERING—SOLIDS

After a recrystallization, you usually collect the new crystals by suction on a Buchner funnel (see Chapter 13, "Recrystallization"). For microscale quantities, you may have to use a Hirsch funnel—a tiny Buchner funnel with sloping sides and a flat, porous plate (see Chapter 6, "Other Interesting Equipment"). Or you might need the high-tech power of Craig tubes (see Chapter 14, "Recrystallization: Microscale"). Or you might be able to get away with a Pasteur pipet (Fig. 7.3).

1. Get the vial or tube containing the ice-cold crystalline mixture and stir it a bit with a Mayo-type cotton-plugged pipet—the cotton plug pushed right to the end of the tip.

2. Slowly—there's that word again—push air out of the pipet while you bring the tip to the bottom of the container.

3. Slowly—there's that word yet again—draw solvent into the pipet, leaving crystals behind.

Now you might want to pack the crystals down a bit by rapping the container on a hard surface. Then repeat the filtration above.

You can wash these crystals with a few drops of cold, fresh solvent and then remove the solvent as many times as is necessary. Eventually, however, you'll have to take the crystals out of the tube and let them dry.

If the tube with crystals has a flattened or even a slightly rounded bottom, you might be able to get away with using an unplugged pipet. The square end of the pipet can fit even a test tube bottom fairly closely, letting solvent in and keeping crystals out. With conical vials, there's more trouble getting a close glass-to-glass contact

I recommend pipet
cotton-plugged to
very tip (Mayo style)

Solvent

Crystals

Solvent

Crystals

Tight
glass-to-glass
seal

FIGURE 7.3 Pasteur pipets to filter solids.

(not impossible, though), and it might be easier to stick to the Mayo-type plugged pipet (Fig. 7.3).

In all cases, make sure you keep the crystallization tube cold and that you wait, even the small length of time it takes, for the pipet tip to cool as well. You don't want the pipet tip warming the solution.

- *Don't fool with these things and stick yourself or another.*

- *Don't use the M.D. injection technique; these are not darts.*

Often, you have to handle samples by syringe. There are about 25 rules for handling syringes. The first five are:

1. These are dangerous—watch it.
2. Watch it—don't stick yourself.
3. Don't stick anyone else.
4. Be careful.
5. Don't fool around with these things.

The other 20 rules—well, you get the idea. If that's not enough, some states require *extremely* tight control of syringes and needles. They might be kept under double lock and key in a locked cabinet in a locked room and require extensive sign-out *and* sign-in procedures. And if you do get stuck, bring yourself and the syringe to the instructor immediately.

You'll probably use a glass or plastic syringe with a **Luer tip** (Fig. 8.1). This is a standard-shape tip that can fit various pieces of apparatus, including the syringe needle. All you do is push the tip into the receptacle. Don't confuse this with the **Luer-Lok,** a special collar placed around the Luer tip designed to hold, well, *lock* the syringe barrel onto the apparatus, including needles (Fig. 8.2). You push the lock onto the receptacle and, with a twist, lock the syringe barrel to it. You'll have to *untwist* this beast to release the syringe barrel.

Plastic syringe

Luer tip fits
syringe needle

Luer tip

Luer tip slides into
Luer lock fittings

FIGURE 8.1 Luer-tip syringe and some of its fittings.

FIGURE 8.2   Luer-Lok syringe and some of its fittings.

The biggest mistake—one that I still make—is losing my patience when filling a syringe with a needle attached. I always tend to pull the plunger quickly, and I wind up sucking air into the barrel. Load the syringe slowly. If you do get air in the barrel, push the liquid back out and try again.

Injecting a sample through a Teflon-lined plastic disk, or a rubber septum, is fairly straightforward. Don't get funny and try slinging the needle—medical school-style—through the disk. Use two hands if you have to: one to hold the barrel and the other to keep the plunger from moving. Slowly push the plunger into the barrel to deliver the liquid. Microliter syringe handling and injection through a septum is also discussed in Chapter 30, "Gas Chromatography."

THE RUBBER SEPTUM

If you're using syringes, you should know about the **rubber septum.** You put the smaller end into the opening you're going to inject into, and then roll the larger end over the *outside* of the opening (Fig. 8.3). You might use a length of copper wire to keep the rubber septum from coming out of the opening, especially if there's any chance of a pressure buildup inside the apparatus.

Roll large end
over the outside
of the joint

Push small end
into joint

Inject here

Wire the septum
on (optional)

FIGURE 8.3 The rubber septum.

CLEAN AND DRY

- *Use elbow grease, too.*
- *Don't dry stuff that'll have water in it.*
- *Overnight drying is usually the best.*

Once you've identified your apparatus, you may find you have to clean it.

1. Wash your glassware at the *end* of the lab day. That way you'll have clean and dry glassware, ready to go for the next lab. This may be difficult to do if you perform an experiment on the day you check in.

2. A *little* solvent, a *little* detergent, and a *lot* of elbow grease. These are the correct proportions for a cleaning solution. You do not need all the soap on the planet, nor do you have to fill the glassware to overflowing with soap solution. *Agitation* is the key here. The more you agitate a small amount of soap solution, the less you agitate your instructor by wasting your time and supplies, and the more effective your cleaning will be.

3. Special Buchner funnel cleaning alert: The standard ceramic Buchner funnel is not transparent, and you can't see whether the bums who used the funnel the last time to collect a highly colored product cleaned the funnel properly. The first time you Buchner-filter crystals from an alcohol solution, the colored impurity dissolves and bleeds up into your previously clean crystals, and you may have to redo your entire experiment. I'd rinse the Buchner funnel with a bit of hot ethanol before I used it, just for insurance.

DRYING YOUR GLASSWARE WHEN YOU DON'T NEED TO

"It's late. Why haven't you started the experiment yet?"

"I washed all my glassware and spent half an hour drying it."

"What technique are we doing?"

"Steam distillation."

"Steam goes through the entire setup, does it?"

A nodding head responds.

"What's condensed steam?"

"Water . . ."

There are all sorts of variations, but they boil down to this: You've taken all this time to dry your glassware only to put water in it. Writers of lab manuals are very tricky about this. Perhaps they say you'll be using *steam.* Or maybe *5% aqueous* sodium bicarbonate solution. Or even that a by-product of your reaction is H_2O. Condense *steam* and you get *what?* An *aqueous* solution has *what* for a solvent? H_2O is *what?*

Look for sources of water other than plain water. If a "water-and" mixture is going to be in the equipment anyway, drying to perfection is silly.

DRYING YOUR GLASSWARE WHEN YOU DO NEED TO

If you wash your glassware *before* you quit for the day, the next time you need it, it'll be clean and dry. There are only a few reactions you might do that need superclean, superdry apparatus, and you should be given special instructions when that's necessary. (In their book *Experimental Organic Chemistry,* 2nd edition, McGraw-Hill, 1986, authors H. D. Durst and G. W. Gokel claim that glassware dried overnight is dry enough for the Grignard reaction—an *extremely* moisture-sensitive reaction—and that flame-drying can be avoided unless the laboratory atmosphere is extremely humid.)

Don't use the compressed air from the compressed air lines in the lab for drying anything. These systems are full of dirt, oil, and moisture from the pumps, and they will get your equipment dirtier than before you washed it.

Yes, there are a few quick ways of drying glassware in case of emergency. You can rinse *very wet* glassware with a small amount of acetone, *drain the glassware very well,* and put the glassware in a drying oven (about 100°C) for a short spell. The acetone washes the water off the glassware very well (the two liquids are **miscible,** that is, they mix in all proportions), and also the liquid left behind is acetone-rich and evaporates faster than water.

If you have one of those "kits" to store your glassware in, make really sure all the acetone is gone. The plastic liners of some of these kits can dissolve in acetone—and there you'll be next time, chipping your glassware out of your kit.

- *Add small amounts until the job is done.*
- *If there's dry drying agent in the flask, the liquid must be dry.*

When you've prepared a liquid product, you must dry the liquid before you finally distill and package it by treating the liquid with a **drying agent.** Drying agents are usually certain anhydrous salts that combine with the water in your product and hold it as **water of crystallization.** When all of the water in your sample is tied up with the salt, you gravity-filter the mixture. The dried liquid passes through the filter paper, and the **hydrated salt** stays behind.

TYPICAL DRYING AGENTS

1. *Anhydrous calcium chloride.* This is a very popular drying agent, inexpensive and rapid, but of late I've become disappointed in its performance. It seems that the calcium chloride powders a bit from storage and abuse, and this **calcium chloride dust** can go right through the filter paper with the liquid. So a caution: If you must use anhydrous calcium chloride, be sure it is granular. There are 4–8-mesh pellets that work fairly well. Avoid powdered calcium chloride, or granular anhydrous calcium chloride that's been around long enough to become pulverized. And don't add to the problem by leaving the lid off the jar of the drying agent; that's the abuse I was talking about.

    Anhydrous calcium chloride tends to form *alcohols of crystallization,* so you really can't use it to dry alcohols.

2. *Anhydrous sodium carbonate and anhydrous potassium carbonate.* These are useful drying agents that are basically **basic.** As they dry your organic compound, any carbonate that dissolves in the tiny amounts of water in your sample can neutralize any tiny amounts of acid that may be left in the liquid. If your product is *supposed* to be acidic (in contrast to being *contaminated* with acid), you should avoid these drying agents.

3. *Anhydrous sodium sulfate.* This is a neutral drying agent that has a high capacity for water, so you can remove huge quantities of water from very wet samples. It's a little slow, and not too efficient, but it is cheap, and it forms a **granular** hydrate from which you can easily decant your dried liquid. It loses water above 32°C, so you might have to pay attention to the solution temperature.

4. *Anhydrous magnesium sulfate.* In my opinion, anhydrous (anh.) $MgSO_4$ is about the best all-around drying agent. It has a drawback, though. Since it's a fine powder, lots of your product can become trapped on the surface. *This is not the same as water of crystallization.* The product is *only on the surface, not inside the crystal structure.* You can wash your product out of the powder with a little fresh solvent.

5. *Drierite.* Drierite, a commercially available brand of anhydrous calcium sulfate, has been around a long time and is a popular drying agent. You can put it in liquids and dry them or pack a drying tube with them to keep the moisture in the air from getting into the reaction setup. But be warned: There is also **blue**

Drierite. This has an indicator, a cobalt salt, that is *blue when dry, pink when wet.* Now, you can easily tell when the drying agent is no good. Just look at it. Unfortunately, this stuff is not cheap, so don't fill your entire drying tube with it just because it'll look pretty. Use a small amount mixed with white Drierite, and when the blue pieces turn pink, change the entire charge in your drying tube. You can take a chance using blue Drierite to dry a liquid directly. Sometimes the cobalt compound dissolves in your product. Then you have to clean and dry your product all over again.

USING A DRYING AGENT

1. Put the liquid or solution to be dried into an Erlenmeyer flask.
2. Add *small* amounts of drying agent and swirl the liquid. Each time you swirl and stop, see if the powered drying agent is taking some time to settle down, and looks like, well, dry drying agent. If the drying agent in the liquid is dry, then the liquid is dry. Larger granulated drying agents might not behave so nicely, and fall rapidly, wet or dry. If you can see that the liquid is no longer cloudy, the water is gone and the liquid is dry.
3. Add just a bit more drying agent and swirl one final time.
4. Gravity-filter through filter paper (see Figure 13.3), or Pasteur pipet (see Figure 7.2).
5. If you've used a carrier solvent, then evaporate or distill it off, whichever is appropriate. Then you'll have your clean, dry product.

FOLLOWING DIRECTIONS AND LOSING PRODUCT ANYWAY

"Add 5 g of anhydrous magnesium sulfate to dry the product." Suppose your yield of product is lower than that in the book. Too much drying agent and not enough product—zap! It's all sucked onto the surface of the drying agent. Bye-bye product. Bye-bye grade.

Always add the drying agent slowly to the product in small amounts.

Now about those small amounts of product (usually liquids):

1. Dissolve your product in a *low-boiling-point solvent.* Maybe ether or hexane or the like. Now dry this whole solution, and gravity-filter. Remove the solvent carefully. Hoo-ha! Dried product.
2. Use chunky dehydrating agents like anhydrous calcium sulfate (Drierite). Chunky drying agents have a much smaller surface area, so not as much of the product is absorbed.

DRYING AGENTS: MICROSCALE

Using drying agents in microscale work follows the same rules as drying on larger scales. Chunky Drierite becomes a problem, and you may have to chop it up a bit (good luck!) before you use it. And, as ever, beware directions that command you to fill a cotton-plugged pipet (see Chapter 7, "Pipet Tips") with exactly 2.76 g of drying agent: Again, this is a *typical* amount of drying agent to be used to dry the *typical* amount of product. The one nice thing about drying your liquids this way is that you can also filter impurities out of the liquid at the same time (see Figure 7.2*b*).

And remember, at microscale it's more important than ever to consider washing the drying agent with fresh solvent to recover as much of your product as possible.

DRYING IN STAGES: THE CAPACITY
AND EFFICIENCY OF DRYING AGENTS

Actually, using drying agents can be a bit more complicated than I've shown, especially if you have to choose the agent on your own. Fact is, if you dried an organic liquid with anhydrous sodium sulfate alone, you could wind up with water still present in the organic liquid, even if it looks dry (not cloudy). Sodium sulfate has a high capacity—it binds a lot of water per gram—but a low efficiency—it leaves water in the organic liquid. So why use it? Sodium sulfate is cheap. Coupled with its high capacity, it's ideal for "predrying" liquids; in a second drying stage, you could use a smaller amount of a lower-capacity, higher-efficiency drying agent.

EXERCISES

1. Phosphorous pentoxide is a heck of a drying agent. In a sealed enclosure, after some time there is no measurable amount of water in the air above the pentoxide. It is also entirely unsuitable for drying many liquids or solutions you would come across in your organic laboratory. What's with that?

2. Anhydrous potassium carbonate is useful for drying amines. Acids, not so much. Why?

3. You're drying a product with anhydrous magnesium sulfate. You're not entirely sure if you added enough. Discuss the phrase "If you see dry drying agent in the flask, it's dry" in this context.

- *Samples clean and dry.*
- *Properly labeled vials, eh?*

The fastest way to lose points is to hand in messy samples. Lots of things can happen to foul up your product. The following are unforgivable sins! Repent and avoid!

SOLID PRODUCT PROBLEMS

1. **Trash in the sample.** Redissolve the sample, gravity-filter, and then evaporate the solvent.
2. **Wet solids.** Press out on filter paper, break up, and let dry. The solid shouldn't stick to the sides of the sample vial. Tacky!
3. **Extremely wet solids (solid floating in water).** Set up a gravity filtration (see Chapter 13, "Recrystallization") and filter the liquid off the solid. Remove the filter paper cone with your solid product, open it up, and leave it to dry. Or remove the solid and dry it on fresh filter paper as above. Use lots of care, though. You don't want filter paper fibers trapped in your solid.

LIQUID PRODUCT PROBLEMS

1. **Water in the sample.** This shows up as droplets or as a layer of water on the top or the bottom of the vial, or *the sample is cloudy.* Dry the sample with a drying agent (see Chapter 10, "Drying Agents"), and gravity-filter into a clean, dry vial.
2. **Trash floating in the sample.** For that matter, it could be on the bottom, lying there. Gravity-filter into a clean, dry vial.
3. **Water in the sample when you don't have a lot of sample.** Since solid drying agents can absorb lots of liquid, what can you do if you have a tiny amount of product to be dried? Add some solvent that has a low boiling point. It must dissolve your product. Now you have a lot of liquid to dry, and *if a little gets lost, it is not all product.* Remove this solvent after you've dried the solution. Be careful if the solvent is flammable. *No flames!*

THE SAMPLE VIAL

Sad to say, an attractive package can sell an inferior product. So why not sell yours? Dress it up in a **neat new label.** Put on:

1. **Your name.** Just in case the sample gets lost on the way to camp.
2. **Product name.** So everyone will know what is in the vial. What does "Product from part C" mean to you? Nothing? Funny, it doesn't mean anything to instructors either.
3. **Melting point (solids only).** This is a range, like "mp 96–98°C" (see Chapter 12, "The Melting-Point Experiment").

4. *Boiling point (liquids only).* This is a range, like "bp 96–98°C" (see Chapter 19, "Distillation").

5. *Yield.* If you weigh the empty vial and cap, you have the **tare.** Now add your product and weigh the full vial. Subtract the *tare* from this *gross weight* to get the **net weight** (yield, in grams) of your product.

6. *Percent yield.* Calculate the percent yield (see Chapter 2, "Keeping a Notebook"), and put it on the label.

You may be asked for more data, but the things listed above are a good start down the road to good technique.

P.S. Gummed labels can fall off vials, and pencil will smear. *Always use waterproof ink!* And a piece of transparent tape over the label will keep it on.

HOLD IT! DON'T TOUCH THAT VIAL

Welcome to "You Bet Your Grade." The secret word is **dissolve.** Say it slowly as you watch the cap liner in some vials dissolve into your nice, clean product and turn it all goopy. This can happen. A good way to prevent it is to cover the vial with aluminum foil before you put the cap on. Just make sure the product does not react with aluminum. Discuss this procedure at length with your instructor.

- *Never remelt samples.*
- *Tiny amounts. If you can just see it, you can see it melt.*

A **melting point** is the temperature at which the first crystal just starts to melt until the temperature at which the last crystal just disappears. Thus the melting point (abbreviated mp) is actually a **melting range.** You should report it as such, even though it is *called* a melting point, for example, mp 147–149°C.

People always read the phrase as melting *point* and never as *melting* point. There is this uncontrollable, driving urge to report one number. No matter how much I've screamed and shouted at people *not* to report one number, they almost always do. It's probably because handbooks list only one number, the upper limit.

Generally, melting points are taken for two reasons.

1. *Determination of purity.* If you take a melting point of your compound and it starts melting at 60°C and doesn't finish until 180°C, you might suspect something is wrong. A melting range *greater than 2°C* usually indicates an impure compound. (As with all rules, there are exceptions. There aren't many to this one, though.)

2. *Identification of unknowns.*
   a. If you have an unknown solid, take a melting point. Many books (ask your instructor) contain tables of melting points and lists of compounds that may have particular melting points. One of them may be your unknown. You may have 123 compounds to choose from. A little difficult, but that's not all the compounds in the world. Who knows? Give it a try. If nothing else, you know the melting point.
   b. Take your unknown, and mix it *thoroughly* with some chemical you think might be your unknown. You might not get a sample of it, but you can ask. Shows you know something. Then,
      (1) If the mixture melts at a *lower* temperature, over a *broad range,* your unknown is not the same compound.
      (2) If the mixture melts at the *same temperature, same range,* it's a good bet it's the *same compound.* Try another run, though, with a different ratio of your unknown and this compound just to be sure. A *lower* melting point with a *sharp range* would be a special case called a **eutectic mixture,** and you, with all the other troubles in lab, just might accidentally hit it. On lab quizzes, this procedure is called

"Taking a mixed melting point."

Actually, "taking a *mixture* melting point," the melting point of a mixture, is more correct. But I have seen this expressed both ways.

SAMPLE PREPARATION

You usually take melting points in thin, closed-end tubes called **capillary tubes.** They are also called **melting-point tubes** or even **melting-point capillaries.** The terms are interchangeable, and I'll use all three.

Sometimes you may get a supply of tubes that are open on *both ends*! You don't just use these as is. Light a burner and, before you start, close off one end. Otherwise your sample will fall out of the tube (see "Closing Off Melting-Point Tubes" later in this chapter).

Take melting points only on *dry, solid* substances, *never* on liquids or solutions of solids *in* liquids or on wet or even damp solids. Only on dry solids!

To help dry damp solids, place the damp solid on a piece of filter paper and fold the paper around the solid. Press. Repeat until the paper doesn't get wet. Yes, you may have to use fresh pieces of paper. Try not to get filter paper fibers in the sample, OK?

Occasionally, you may be tempted to dry solid samples in an oven. *Don't*— unless you are specifically instructed to. I know some students who have decomposed their products in ovens and under heat lamps. With the time they save quickly decomposing their product, they can repeat the entire experiment.

Loading the Melting-Point Tube

Place a small amount of *dry* solid on a new filter paper (Fig. 12.1). Thrust the open end of the capillary tube into the middle of the pile of material. Some solid should be trapped in the tube. *Carefully* turn the tube over, taking care not to let the solid fall

FIGURE 12.1 Loading a melting-point tube.

FIGURE 12.2   Closing off an mp tube with a flame.

back out of the tube, until the closed end is down. Remove any solid sticking to the outside. Now, the solid must be packed down. If you have trouble with gouging the filter paper, you might put the solid on a watch glass instead of filter paper.

Traditionally, the capillary tube, turned upright with the *open end up,* is stroked with a file or trapped on the bench top. Unless they are done *carefully,* these operations *may break the tube.* A safer method is to drop the tube, *closed end down,* through a length of glass tubing. You can even use your condenser or distilling column for this purpose. When the capillary strikes the bench top, the compound will be forced into the closed end. You may have to do this several times. If there is not enough material in the mp tube, thrust the open end of the tube into the mound of material and pack it down again. Use your own judgment; consult your instructor.

Use the smallest amount of material that can be seen to melt.

Closing Off Melting-Point Tubes

If you have melting-point tubes that are *open at both ends* and you try to take a melting point with one, it should come as no surprise when your compound falls out of the tube. You'll have to *close off one end* to keep your sample from falling out (Fig. 12.2). So light a burner, and get a "stiff" small blue flame. Slowly touch the end of the tube to the side of the flame, and hold it there. You should get a yellow sodium flame, and the tube will close up. There is no need to rotate the tube. And remember, *touch—just touch*—the edge of the flame, and hold the tube there. Don't feel you have to push the tube way into the flame.

MELTING-POINT HINTS

1. Use only the smallest amount that you can see melt. Larger samples will heat unevenly.
2. Pack down the material as much as you can. Left loose, the stuff will heat unevenly.
3. Never remelt any sample. It may undergo nasty chemical changes such as oxidation, rearrangement, or decomposition.

4. Make up more than one sample. One is easy, two is easier. If something goes wrong with one, you have another. Duplicate and even triplicate runs are common.

THE MEL-TEMP APPARATUS

The Mel-Temp apparatus (Fig. 12.3) substitutes for the Thiele tube or open beaker and hot oil methods (see "Using the Thiele Tube" later in this chapter). Before you use the apparatus, there are a few things you should look for.

FIGURE 12.3   The Mel-Temp apparatus.

Thermometer

mp tube slot

Heating block

Eyepiece

Light source

Typical view in eyepiece

mp tubes in two of three channels

FIGURE 12.4 Close-up of the viewing system for the Mel-Temp apparatus.

1. **Line cord.** Brings AC power to the unit. Should be plugged into a live wall socket. (See J. E. Leonard and L. E. Mohrmann, *J. Chem. Educ.*, **57,** 119 (1980) for a modification in the wiring of older units to make them less lethal. It seems that, even with the three-prong plug, there can still be a shock hazard. *Make sure your instructor knows about this!*)

2. **On–off switch.** Turns the unit on or off.

3. **Fuse.** Provides electrical protection for the unit.

4. **Voltage control.** Controls the *rate* of heating, *not the temperature!* The higher the setting, the faster the temperature rises.

5. **Light source.** Provides illumination for samples.

6. **Eyepiece.** Magnifies the sample (Fig. 12.4).

7. **Thermometer.** Gives the temperature of the sample, and upsets the digestion when you're not careful and you snap it off in the holder.

OPERATION OF THE MEL-TEMP APPARATUS

1. *Imagine yourself getting burned if you're not careful.* Never assume that the unit is cold. Read the temperature, and wait for the heating block to cool if the temperature is fewer than 20°C below the melting point of your compound. You'll have to wait for the block to cool to room temperature if you have no idea of the approximate melting point of your compound.

2. Place the loaded mp tube in one of the three channels in the opening at the top of the unit (Fig. 12.4).

3. Set the voltage control to zero if necessary. There are discourteous folk who do not reset the control when they finish using the equipment.

4. Turn the on–off switch to ON. The light source should illuminate the sample. If not, call for help.

5. Now science turns into art. Set the *voltage control* to *any* convenient setting. The point is to get up to *within 20°C* of the *supposed* melting point. Yep, that's right. If you have no idea what the melting point is, it may require several runs as you keep skipping past the points with a temperature rise of 5–10°C per minute. A convenient setting is *40*. This is just a suggestion, not an article of faith.

6. After you've melted a sample, *throw it away!*

7. Once you have an idea of the melting point (or looked it up in a handbook, or were told), *get a fresh sample,* and bring the temperature up quickly at about *5–10°C per minute to within 20°C* of this approximate melting point. Then, turn down the *voltage control* to get a *2°C per minute rise.* Patience!

8. When the first crystals *just start to melt,* record the temperature. When the *last crystal just disappears,* record the temperature. If both points appear to be the same, either the sample is extremely pure or the temperature rise was *too fast.*

9. Turn the on–off switch to OFF. You can set the voltage control to zero for the next person.

10. Remove all capillary tubes.

Never use a wet rag or sponge to quickly cool off the heating block. This might permanently warp the block. You can use a cold metal block to cool it if you're in a hurry. Careful. If you slip, you may burn yourself.

THE FISHER-JOHNS APPARATUS

The Fisher-Johns apparatus (Fig. 12.5) is different in that you don't use capillary tubes to hold the sample. Instead, you sandwich your sample between two round microscope cover slips (thin windows of glass) on a heating block. This type of melting-point apparatus is called a **hot stage.** It comes complete with spotlight. Look for the following.

1. *Line cord* (at the back). Brings AC power to the unit. Should be plugged into a live wall socket.

2. *On–off switch.* Turns the unit on or off.

3. *Fuse* (also at the back). Provides electrical protection for the unit.

FIGURE 12.5 The Fisher-Johns apparatus.

4. **Voltage control.** Controls the *rate* of heating, *not the temperature*! The higher the setting, the faster the temperature rise.

5. **Stage light.** Provides illumination for samples.

6. **Eyepiece.** Magnifies the sample.

7. **Thermometer.** Gives the temperature of the sample.

8. **Thermometer end cap.** Keeps the thermometer from falling out. If the cap becomes loose, the thermometer tends to go belly-up and the markings turn over. Don't try to fix this while the unit is hot. Let it cool so you won't get burned.

9. **The hot stage.** This is the heating block on which samples are melted.

OPERATION OF THE
FISHER-JOHNS APPARATUS

1. *Don't assume that the unit is cold.* That is a good way to get burned. Read the temperature, and wait for the heating block to cool if the temperature is fewer than 20°C below the melting point of your compound. You'll have to wait until the block is at room temperature if you have no idea of the approximate melting point of your compound.

2. Keep your grubby fingers off the cover slips. Use tweezers or forceps.

3. Place a clean round glass cover slide in the well on the hot stage. *Never melt any samples directly on the metal stage.* Ever!

4. Put a few crystals on the glass. Not too many. As long as you can see them melt, you're all right.

5. Put another cover slide on top of the crystals to make a sandwich.

6. Set the voltage control to zero, if it's not already there.

7. Turn the on–off switch to ON. The light source should illuminate the sample. If not, call for help!

8. Now science turns into art. Set the *voltage control to any* convenient setting. The point is to get up to *within 20°C* of the *supposed* melting point. Yep, that's right. If you have no idea of the melting point, it may require several runs as you keep skipping past the point with a temperature rise of 5–10°C per minute. A convenient setting is *40*. This is just a suggestion, not an article of faith.

9. After you've melted a sample, let it cool, and remove the sandwich of sample and cover slides. *Throw it away!* Use an appropriate waste container.

10. Once you have an idea of the melting point (or have looked it up in a hand-book, or you were told), *get a fresh sample,* and bring the temperature up quickly at about *5–10°C per minute to within 20°C* of this approximate melt-ing point. Then turn down the *voltage control* to get a *2°C-per-minute rise.* Patience!

11. When the first crystals *just start to melt,* record the temperature. When the last crystal *just disappears,* record the temperature. If both points appear to be the same, either the sample is extremely pure or the temperature rise was *too fast.*

12. Turn the on–off switch to OFF. Now set the voltage control to zero.

13. Let the stage cool, then remove the sandwich.

THE THOMAS-HOOVER APPARATUS

The Thomas-Hoover apparatus (Fig. 12.6) is the electromechanical equivalent of the Thiele tube or open beaker and hot oil methods (see "Using the Thiele Tube" later in this chapter). It has lots of features, and you should look for the following.

1. *Light box.* At the top of the device toward the back, a box holds a fluorescent lightbulb behind the thermometer. On the right side of this box are the fluores-cent light switches.

    I'm told that there are versions of this beast that don't have a light box. I could write another whole set of instructions for that type, or, if you have such a unit, you could try ignoring anything that pertains to the light box.

2. *Fluorescent light switches.* Two buttons. Press and hold the red button down for a bit to light the lamp; press the black button to turn the lamp off.

3. *Thermometer.* A special 300°C thermometer in a metal jacket is immersed in the oil bath that's in the lower part of the apparatus. Two slots have been cut in the jacket to let light illuminate the thermometer scale from behind and to let a thermometer periscope read the thermometer scale from the front.

4. *Thermometer periscope.* In front of the thermometer, this periscope lets you read a small magnified section of the thermometer scale. By turning the small knob at the lower right of this assembly, you track the movement of the mercury

Fluorescent light in this box

Thermometer in metal casing

Holder for melting point tubes

Samples fit in holes here (really!)

Cords to fluorescent light and heater

Thermometer reader periscope

Fluorescent light ON button (red)

Fluorescent light OFF button (black)

Thermometer reader periscope knob

Read temperature here in mirror

Adjustable magnifying glass (samples in window behind it)

Temperature control

Vibrator on–off switch

On–off power switch

Stirrer motor control

LOW   HIGH

FIGURE 12.6 The Thomas-Hoover apparatus.

thread, and an image of the thread and temperature scale appears in a stationary mirror just above the sample viewing area.

5. *Sample viewing area.* A circular opening is cut in the front of the metal case so that you can see your samples in their capillary tubes (and the thermometer bulb), all bathed in the oil bath. You put the tubes into the oil bath through the holes in the capillary tube stage.

6. *Capillary tube stage.* In a semicircle around the bottom of the jacketed thermometer, yet behind the thermometer periscope, are five holes through which you can put your melting-point capillaries.

7. *Heat.* Controls the rate of heating, not the temperature. The higher the setting, the faster the temperature rise. At Hudson Valley Community College we've had a stop put in, and you can turn the dial only as far as the number 7. When it gets up to 10, you always smoke the oil. Don't do that.

8. *Power on–off switch.* Turns the unit on or off.

9. *Stirrer motor control.* Sets the speed of the stirrer from low to high.

10. *Vibrator on–off switch.* Turns the vibrator on or off. It's a spring-return switch, so you must hold the switch in the ON position. Let go, and it snaps to OFF.

11. *Line cords.* One cord brings AC power to the heater, stirrer, sample light, and vibrator. The other cord brings power to the fluorescent light behind the thermometer. Be sure that both cords are plugged into live wall sockets.

OPERATION OF THE THOMAS-HOOVER APPARATUS

1. If the fluorescent light for the thermometer is not lit, press the red button at the right side of the light box, and hold it down for a bit to start the lamp. The lamp should remain lit after you release the button.

2. Look in the thermometer periscope, turn the small knob at the lower right of the periscope base, and adjust the periscope to find the top of the mercury thread in the thermometer. Read the temperature. Wait for the oil bath to cool if the temperature is fewer than 20°C below the approximate melting point of your compound. You'll have to wait until the bath is at room temperature if you have no idea of the melting point. You don't want to plunge your sample into oil that is so hot it might melt too quickly or melt at an incorrect temperature.

3. Turn the voltage control to zero, if it isn't there already.

4. Turn the power on–off switch to ON. The oil bath should become illuminated.

5. Insert your capillary tube in one of the capillary tube openings in the capillary tube stage. *This is not simple.* Be careful. If you snap a tube at this point, the entire unit may have to be taken apart to remove the pieces. You have to angle the tube toward the center opening and angle the tube toward you (as you face the instrument) at the same time (Fig. 12.7). The capillary tube has to negotiate a second set of holes, so you have to angle the tube toward the center.

6. Adjust the magnifying glass for the best view of your sample.

7. Turn the stirrer knob so that the mark on the knob is about halfway between the slow and fast markings on the front panel. That's just a suggestion. I don't have any compelling reasons for it.

Sample tubes fit in a conical arrangement (exaggerated here) around the capillary tube stage

Thermometer periscope removed for clarity

mp tube with sample

Oil-filled beaker behind window

Heating element

FIGURE 12.7 Close-up of the viewing system for the Thomas-Hoover apparatus.

8. Adjust the thermometer periscope to give you a good view of the top of the mercury thread in the thermometer.

9. Now science turns into art. Set the *heat control* to *any* convenient setting. The point is to get up to *within 20°C* of the *supposed* melting point. If you have no idea of the melting point, it may require several runs as you keep skipping past the point with a temperature rise of 5–10°C per minute. A convenient setting is *4.* This is just a suggestion, not an article of faith.

10. Remember, you'll have to keep adjusting the thermometer periscope to keep the top of the mercury thread centered in the image.

11. After you've melted a sample, *throw it away!*

12. Once you have an idea of the melting point (or looked it up in a handbook, or were told), *get a fresh sample* and bring the temperature up quickly at about *5–10°C per minute to within 20°C* of this approximate melting point. Then turn down the *heat control* to get a *2°C-per-minute rise.* Patience!

13. When the first crystals *just start to melt,* record the temperature. When the *last crystal just disappears,* record the temperature. If both points appear to be the same, either the sample is extremely pure or the temperature rise was *too fast.* If you record the temperature with the horizontal index line in the mirror matched to the lines etched on both sides of the periscope window and the top of the mercury thread at the same time, you'll be looking at the thermometer scale head-on. This will give you the smallest error in *reading* the temperature (Fig. 12.8).

14. Don't turn the control much past 7. You can get a bit beyond 250°C at that setting, and that should be *plenty* for any solid compound that you might

FIGURE 12.8 Reading the temperature.

prepare in organic lab. Above this setting, there's a real danger of smoking the oil.

15. Turn the power switch to OFF. You can also set the *heat control* to zero for the next person.

16. Press the black button on the right side of the light box, and turn the fluorescent light off.

17. Remove all capillary tubes and dispose of them properly.

There are a few more electric melting-point apparatuses around, and most of them work in the same way. A **sample holder, magnifying eyepiece,** and **voltage control** are common; an apparently essential feature of these devices is that dial markings are almost *never* temperature settings. That is, a setting of **60** will not give a temperature of 60°C but probably one that is much higher.

USING THE THIELE TUBE

With the Thiele tube (Fig. 12.9), you use hot oil to transfer heat evenly to your sample in a melting-point capillary, just like the metal block of the Mel-Temp apparatus does. You heat the oil in the side arm and it expands. The hot oil goes up the side arm, warming your sample and thermometer as it touches them. Now the oil is cooler, and it falls to the bottom of the tube, where it is heated again by a burner. This cycle goes on automatically as you do the melting-point experiment in the Thiele tube.

FIGURE 12.9 Taking melting points with the Thiele tube.

Don't get any water in the tube, or when you heat the tube the water can boil and throw hot oil out at you. Let's start from the beginning.

Cleaning the Tube

This is a bit tricky, so don't do it unless your instructor says so. Also, check with your instructor *before* you put fresh oil in the tube.

1. Pour the old oil out into an appropriate container, and let the tube drain.

2. Use a hydrocarbon solvent (hexane, ligroin, petroleum ether—and *no flames!*) to dissolve the oil that's left.

3. Get out the old soap and water and elbow grease, clean the tube, and rinse it out really well.

4. Dry the tube thoroughly in a drying oven (usually >100°C). Carefully take it out of the oven, and let it cool.

5. Let your instructor examine the tube. If you get the OK, *then* add some fresh oil. Watch it. First, *no water.* Second, don't overfill the tube. Normally, the oil expands as you heat the tube. If you've overfilled the tube, oil will crawl out and get you.

Getting the Sample Ready

Here you use a loaded melting-point capillary tube (see "Loading the Melting-Point Tube" earlier in the chapter) and attach it directly to the thermometer. Unfortunately, the thermometer has bulges; there are some problems, and you may snap the tube while attaching it to the thermometer.

1. Get a thin rubber ring or cut one from a piece of rubber tubing.

2. Put the *bottom* of the loaded mp tube *just above* the place where the thermometer constricts (Fig. 12.10), and carefully roll the rubber ring onto the mp tube.

Hold mp tube on straight part

Move ring up and over tube

Slide mp tube down

Move ring up

No pressure here! (Tube may snap!)

Ring near top of tube

Final resting place!

Middle of sample at middle of bulb

FIGURE 12.10   Attaching the mp tube to the thermometer without a disaster.

3. Reposition the tube so that the sample is near the center of the bulb and the rubber ring is near the open end. *Make sure the tube is vertical.*

4. Roll the rubber ring up to near the top of the mp tube, out of harm's way.

Dunking the Melting-Point Tube

There are more ways of keeping the thermometer suspended in the oil than I care to list. You can cut or file a notch on the side of the cork, drill a hole, and insert the thermometer (*be careful!*). Finally, cap the Thiele tube and dunk the mp tube (Fig. 12.9). The notch is there so that pressure will not build up as the tube is heated. *Keep the notch open, or the setup may explode.*

But this requires drilling or boring corks, something you try to avoid. (Why have ground-glass jointware in the undergraduate lab?) You can *gently* hold a thermometer and a cork in a clamp (Fig. 12.11). Not too much pressure, though!

Undrilled cork helps hold thermometer (careful)

Quasi-legal use of thermometer adapter to hold thermometer

Thiele tubes completely open and safe

FIGURE 12.11 Safely suspended thermometer with Thiele tube.

Finally, you might put the thermometer in the **thermometer adapter** and suspend that, clamped gently by the rubber part of the adapter, not by the ground-glass end. Clamping ground glass will score the joint.

Heating the Sample

The appropriately clamped thermometer is set up in the Thiele tube as in Fig. 12.9. Look at this figure *now,* and remember to heat the tube *carefully—always carefully— at the elbow. Then,*

1. If you don't know the melting point of the sample, heat the oil fairly quickly, *but no more than 10°C per minute,* to get a rough melting point. And it will be rough indeed, since the temperature of the thermometer usually lags that of the sample.

2. After this sample has melted, lift the thermometer and attached sample tube *carefully (it may be HOT)* by the thermometer up at the clamp, until they are *just out of the oil.* This way the thermometer and sample can cool, and the hot oil can drain off. Wait for the thermometer to cool to about room temperature before you remove it entirely from the tube. Wipe off some of the oil, reload a melting-point tube (*never remelt melted samples*), and try again. And heat at 2°C per minute this time.

EXERCISES

1. "If you can see it, you can see it melt." This should put the correct amount of a solid in a melting-point tube. If someone has put about one-quarter of an inch of solid in the tube, should they go for an eye exam?

2. Define "melting point."

3. How fast should a melting-point capillary sample be heated?

4. Why not remelt samples? Don't they just cool and solidify in exactly the reverse manner that they heated and melted?

5. Define "freezing point."

RECRYSTALLIZATION

CHAPTER

- *Too much solvent and your crystals don't come back.*

The essence of a recrystallization is **purification.** Messy, dirty compounds are cleaned up, purified, and can then hold their heads up in public again. The sequence of events you use will depend a lot on how messy your crude product is and on just how soluble it is in various solvents.

In any case, you'll have to remember a few things.

1. Find a solvent that will *dissolve the solid while hot.*
2. The same solvent *should not dissolve it while cold.*
3. The *cold solvent* must keep impurities dissolved in it *forever or longer.*

This is the major problem. And it requires some experimentation. That's right! Once again, art over science. Usually, you'll know what you should have prepared, so the task is easier. It requires a trip to **your notebook** and, possibly, a **handbook** (see Chapter 2, "Keeping a Notebook," and Chapter 3, "Interpreting a Handbook"). You have the data on the solubility of the compound in your notebook. What's that, you say? *You don't have the data in your notebook?* Congratulations, you get the highest F in the course.

Information in the notebook (which came from a handbook) for your compound might say, for alcohol (meaning *ethyl* alcohol), **s.h.** Since this means **s**oluble in **h**ot alcohol, it implies that it is **i**nsoluble (**i.**) in cold alcohol. Then alcohol is probably a good solvent for recrystallization of that compound. Also, check on the **color** or **crystalline form.** This is important since:

1. A color in a supposedly white product is an impurity.
2. A color in a colored product is *not* an impurity.
3. The *wrong color* in a product is an impurity.

You can usually assume that impurities are present in small amounts. Then you don't have to guess what possible impurities might be present or what they might be soluble or insoluble in. If your sample is really dirty, the assumption can be fatal. This doesn't usually happen in an undergraduate lab, but you should be aware of it.

FINDING A GOOD SOLVENT

If the solubility data for your compound are not in handbooks, then:

1. Place 0.10 g of your solid (weighed to 0.01 g) in a test tube.
2. Add 3 mL of a solvent, stopper the tube, and shake the bejesus out of it. If *all of the solid dissolves at room temperature,* then your solid is **soluble.** Do *not* use this solvent as a recrystallization solvent. (You must make note of this in your notebook, though.)
3. If none (or very little) of the solid dissolved at room temperature, unstopper the tube and heat it (*careful—no flames and get a boiling stone!*) and

shake it and heat it and shake it. You may have to heat the solvent to a gentle boil. (*Careful!* Solvents with low boiling points often boil away.) If it does *not* dissolve at all, then do not use this as a recrystallization solvent.

4. If the sample *dissolved when hot,* and *did not dissolve at room temperature,* you're on the trail of a good recrystallization solvent. One last test.

5. Place the test tube (no longer hot) in an ice-water bath, and cool it about 5°C or so. If lots of crystals come out, this is good, and this is your recrystallization solvent.

6. Suppose your crystals don't come back when you cool the solution. Get a glass rod into the test tube, stir the solution, rub the inside of the tube with the glass rod, and agitate that solution. If crystals still don't come back, perhaps you'd better find another solvent.

7. Suppose, after all this, you still haven't found a solvent. Look again. Perhaps your compound *completely* dissolved in ethanol at room temperature and would *not* dissolve in water. Aha! Ethanol and water are **miscible** (i.e., they mix in all proportions). You will have to perform a **mixed-solvent recrystallization** (see "Working with a Mixed-Solvent System" later in this chapter).

GENERAL GUIDELINES FOR A RECRYSTALLIZATION

Here are some general rules to follow for purifying any solid compound.

1. Put the solid in an Erlenmeyer flask, not a beaker. If you recrystallize compounds in beakers, you may find the solid climbing the walls of the beaker to get at you as a reminder. A 125-mL Erlenmeyer usually works. Your solid should look comfortable in it, neither cramped nor with too much space. You probably shouldn't fill the flask more than one-fifth to one-fourth full.

2. Heat a large quantity of a proven solvent (see preceding section) to the boiling point, and *slowly add the hot solvent* to the sample in the Erlenmeyer. *Slowly!* A word about solvents: *Fire! Solvents burn! No flames!* A hot plate here would be better. You can even heat solvents in a *steam or water bath.* But—*no flames!*

3. Carefully add the hot solvent to the solid to just dissolve it. In parts. Your first shot of solvent *must* not dissolve all of the solid. Or else you might have gone too far and have more solvent in that flask than to just dissolve your solid. This can be tricky, since hot solvents evaporate, cool down, and so on. Ask your instructor.

4. Add a slight excess of the hot solvent to keep the solid dissolved.

5. If the solution is only slightly colored, the impurities will stay in solution. Otherwise, the big gun, **activated charcoal,** may be needed (see "Activated Charcoal" later in this chapter). Remember, if you were working with a colored compound, it would be silly to try to get rid of all the color, since you would get rid of all the compound and probably all your grade.

6. Keep the solvent hot (*not boiling*) and look carefully to see if there is any trash in the sample. This could be old boiling stones, sand, floor sweepings, and so on. Nothing you'd want to bring home to meet the folks. *Don't confuse real trash with undissolved good product!* If you add more hot solvent, good product will dissolve, and trash will not. If you have trash in the sample, do a **gravity filtration** (see following section).

7. Let the Erlenmeyer flask and the hot solution cool. Slow cooling gives better crystals. Garbage doesn't get trapped in them. But this can take what seems to be an interminable length of time. (I know, the entire lab seems to take an interminable length of time.) So, after the flask cools and it's just *warm* to the touch, then put the flask in an ice-water bath to cool. *Watch it!* The flasks have a habit of turning over in the water baths and letting all sorts of water destroy all your hard work! Also, a really hot flask will shatter if plunged into the ice bath, so again, watch it.

    Why everyone insists on loading up a bucket with ice and trying to force a flask into this mess, I'll never know. How much cooling do you think you're going to get with just a few small areas of the flask barely touching ice?

    Ice bath really means ice-water bath.

8. **"Is it cool enough yet?"** Hey, cool it, yourself. When the solution gets down to the temperature of the ice-water bath (and you should know what that is, eh?), then filter the crystals on a **Buchner funnel** or **Hirsch funnel.**

9. Dry them and take a melting point, as described in Chapter 12, "The Melting-Point Experiment."

GRAVITY FILTRATION

If you find yourself with a flask full of hot solvent and your product dissolved in it, along with all sorts of trash, this is for you. You'll need more hot solvent, a ring stand with a ring attached, possibly a clay triangle, some filter paper, a clean, dry flask, and a **stemless funnel.** Here's how **gravity filtration** works.

1. Fold up a **filter cone** from a piece of filter paper (Fig. 13.1). It should fit nicely, within a single centimeter or so of the top of the funnel. For those who wish to filter with more panache, try using **fluted filter paper** (see "Finishing the final fluted fan," Fig. 13.12).

First fold — Second fold — Open to a cone

FIGURE 13.1 Folding filter paper for gravity filtration.

2. Get yourself a **stemless funnel** or, at least, a **short-stem funnel,** or even a **powder funnel.** Why? Go ahead and *use* a long-stem funnel and watch the crystals come out in the stem as the solution cools, blocking up the funnel (Fig. 13.2).

3. Put the **filter paper cone** in the **stemless funnel.**

4. Support this in a ring attached to a ring stand (Fig. 13.3). If the funnel is too small and you think it could fall through the ring, you may be able to get a **wire** or **clay triangle** to support the funnel in the ring (Fig. 13.4).

5. Put the new, *clean, dry flask* under the funnel to catch the hot solution as it comes through. All set?

6. Get that flask with the solvent, product, and trash hot again. (*No flames!*) You should get some fresh, clean solvent hot as well. (*No flames!*)

7. Carefully pour the hot solution into the funnel. As it is, some solvents evaporate so quickly that product will probably come out on the filter paper. It is often hard to tell the product from the insoluble trash. Then—

8. Wash the filter paper down with *a little hot solvent.* The product will redissolve. The trash won't.

Crystals block stem

FIGURE 13.2 The "too long a funnel stem—oops!" problem.

Hot solvent containing insoluble material

Short-stem funnel

Ring for support

Filter paper cone

Air space

Clean Erlenmeyer flask

Solution with no insoluble impurities

FIGURE 13.3 The gravity filtration setup with aC funnel that fits the iron ring.

9. Now, you let the *trash-free* solution cool, and clean crystals should come out. Since you have probably added solvent to the solution, *don't be surprised if no crystals come out of solution. Don't panic, either!* Just boil away some of the solvent (not out into the room air, please), let your solution cool, and wait for the crystals again. If they *still* don't come back, just repeat the boiling.

Do not boil to dryness!

FIGURE 13.4 A wire triangle holding a small funnel in a large iron ring for gravity filtration.

Somehow, lots of folks think recrystallization means dissolving the solid, then boiling away all the solvent to dryness. No! There must be a way to convince these lost souls that *the impurities will deposit on the crystals.* After the solution has cooled, crystals come out, sit on the bottom of the flask, and *must be covered by solvent!* Enough solvent to keep those nasty impurities dissolved and off the crystals.

THE BUCHNER FUNNEL AND FILTER FLASK

The **Buchner funnel** (Fig. 13.5) is used primarily for separating crystals of product from the liquid above them. If you have been *boiling your recrystallization solvents dry, you should be horsewhipped* and forced to reread these sections on recrystallization!

1. Get a piece of filter paper large enough to cover all of the holes in the bottom plate, and yet *not* curl up the sides of the funnel. It is placed *flat* on the plate (Fig. 13.5).

2. Clamp a **filter flask** to a ring stand. This filter flask, often called a **suction flask,** is a very heavy-walled flask that has a side arm on the neck. A piece of heavy-walled tubing connects this flask to the **water trap** (see Figure 13.6).

Buchner funnel (Top view)

Paper covers all holes but does **not** come up the sides

Crystals in solvent

Crystals
Filter paper
Porous plate
Stopper or rubber adapter
Clamp here
To vacuum trap
Clean filter flask
Ring stand for support

FIGURE 13.5 The Buchner funnel at home and at work.

FIGURE 13.6   A couple of water traps hanging around.

3. Now use a **rubber stopper** or **filter adapter** to stick the Buchner funnel into the top of the filter flask. The Buchner funnel makes the setup top-heavy and prone to be prone—and broken. Clamp the flask first, or go get a new Buchner funnel to replace the one you'll otherwise break.

4. The **water trap** is in turn connected to a source of vacuum, most likely a **water aspirator** (see Figure 13.7).

5. The faucet on the **water aspirator** should be turned on *full blast!* This should suck down the filter paper, which you now *wet with some of the cold recrystallization solvent.* This will make the paper stick to the plate. You may have to push down on the Buchner funnel a bit to get a good seal between the rubber adapter and the funnel.

6. Swirl and pour the crystals and solvent *slowly and directly into the center of the filter paper,* as if to build a small mound of product there. *Slowly!* Don't flood the funnel by filling it right to the brim and waiting for the level to go down. If you do that, the paper may float up, ruining the whole setup.

FIGURE 13.7 A water aspirator.

7. Use a very small amount of the same *cold* recrystallization solvent and a spatula to remove any crystals left in the flask. Then you can use some of the *fresh, cold recrystallization solvent* and slowly pour it over the crystals to wash away any old recrystallization solvent and dissolved impurities.

8. Leave the aspirator on, and let air pass through the crystals to help them dry. You can put a thin rubber sheet, a **rubber dam,** over the funnel. The vacuum pulls it in, and the crystals are pressed clean and dry. You won't have air or moisture blowing through, and possibly decomposing, your product. Rubber dams are neat.

9. When the crystals are dry *and you have a* **water trap,** just turn off the water aspirator. Water won't back up into your flask. (If you've been foolhardy and have filtered without a water trap, just remove the rubber tube connected to the filter flask side arm [Fig. 13.5].)

10. At this point, you may have a *cake of crystals* in your **Buchner funnel.** The easiest way to handle this is to *carefully lift the cake* of crystals out of the funnel *along with the filter paper,* plop the whole thing onto a larger piece of filter paper, and let everything dry overnight. If you are pressed for time, *scrape the damp filter cake from the filter paper, but don't scrape any filter paper fibers into the crystals.* Repeatedly press the crystals between dry sheets of

filter paper, and change sheets until the crystals no longer show any solvent spot after pressing. Those of you who use **heat lamps** may find your white crystalline product turning into instant charred remains.

11. When your cake is *completely dry*, weigh a vial, put in the product, and weigh the vial again. Subtracting the weight of the vial from the weight of the vial and sample gives the weight of the product. This **weighing by difference** is easier and less messy than weighing the crystals directly on the balance. This weight should be included in the label on your **product vial** (see Chapter 5, "Microscale Jointware," and Chapter 11, "On Products").

Just a Note

I've said that a Buchner funnel is used primarily for separating crystals of product from the liquid above them. And in the section on drying agents, I tell people to use a gravity filtration setup to separate a drying agent from a liquid product. Recently, I've had some people get the notion that you can Buchner-filter liquid products from drying agents. I don't advise that. You will probably lose a lot of your product, especially if it has a low boiling point ($<100°C$). Under this *vacuum filtration,* your product simply evaporates along with your grade.

THE HIRSCH FUNNEL AND FRIENDS

You can think of the **Hirsch funnel** (Fig. 13.8) as a smaller, sloping-sided Buchner funnel, and, for the most part, you use it in the same way.

FIGURE 13.8 The Hirsch funnel and friends.

ACTIVATED CHARCOAL

Activated charcoal is ultrafinely divided carbon with lots of places to suck up big, huge, polar, colored impurity molecules. Unfortunately, if you use too much, it'll suck up *your product!* And, if your product was white or yellow, it'll have a funny gray color from the excess charcoal. Sometimes the impurities are untouched and only the product is absorbed. Again, it's a matter of trial and error. Try not to use too much. Suppose you've got a *hot solution* of some solid, *and the solution is highly colored.* Well,

1. First, *make sure your product should not be colored!*

2. Take the flask with your filthy product off the heat, and swirl the flask. This dissipates any superheated areas so that when you add the activated charcoal, the solution doesn't foam out of the flask and onto your shoes.

3. *Add the activated charcoal.* Put a small amount, about the size of a pea, on your spatula; then throw the charcoal in. Stir. The solution should turn black. Stir and heat.

4. Set up the **gravity filtration** and filter off the carbon. It is especially important to *wash off any product caught on the charcoal,* and it is really hard to see anything here. You should take advantage of **fluted filter paper.** It should give a more efficient filtration.

5. Yes, have some extra fresh solvent heated as well. You'll need to add a few milliliters of this to the hot solution to help keep the crystals from coming out on the filter paper. And you'll need more to help wash the crystals off the paper when they come out on it anyway.

6. The filtered solution should be *much cleaner* than the original solution. If not, you'll have to *add charcoal and filter again.* There is a point of diminishing returns, however, and one or two treatments is usually all you should do. Get some guidance from your instructor.

Your solid products should not be gray. Liquid products (yes, you can do liquids!) will let you know that you didn't get all the charcoal out. Often, you can't see charcoal contamination in liquids while you're working with them. The particles stay suspended for awhile, but after a few days, you can see a layer of charcoal on the bottom of the container. Sneaky, those liquids. By the time the instructor gets to grade all the products—*voilà*—the charcoal has appeared.

THE WATER ASPIRATOR: A VACUUM SOURCE

Sometimes you'll need a vacuum for special work like **vacuum distillation** and **vacuum filtration** as with the Buchner funnel. The **water aspirator** is an inexpensive source of vacuum (Fig. 13.7).

When you turn the water on, the water flow draws air in from the side port on the aspirator. The faster the water goes through, the faster the air is drawn in. Pretty neat, huh? I've shown a plastic aspirator, but many of the older metal varieties are still around.

You may have to pretest some aspirators before you find one that will work well. It'll depend on the water pressure in the pipes, too. Even the number of people using aspirators on the same water line can affect the performance of these devices. You can test them by going to an aspirator and turning the faucet on *full blast.* It does help to have a sink under the aspirator. If water leaks out the side port, *tell your instructor and find another aspirator.* Wet your finger and place it over the hole in the side port to feel if there is any vacuum. If there is *no vacuum,* tell your instructor and find another aspirator. Some of these old, wheezing aspirators produce a very weak vacuum. You must decide for yourself if the suction is "strong enough." There should be a **splash guard** or rubber tubing leading the water stream directly into the sink. This will keep the water from going all over the room. If you check and don't find such protection, see your instructor. All you have to do with a fully tested and satisfactory aspirator is to hook it up to the **water trap.**

THE WATER TRAP

Every year I run a chem lab, and when someone is doing a **vacuum filtration,** suddenly I'll hear a scream and a moan of anguish, as water backs up into someone's filtration system. Usually there's not much damage, since the filtrate in the suction flask is generally thrown out. For **vacuum distillations,** however, this **suck-back** is disaster. It happens whenever there's a pressure drop on the water line big enough to cause the flow to decrease so that there is a *greater vacuum in the system than in the aspirator.* Water, being water, flows into the system. Disaster.

So, for your own protection, make up a **water trap** from some stoppers, rubber tubing, a thick-walled Erlenmeyer or filter flask, and a screw clamp (Fig. 13.6). *Do not use garden-variety Erlenmeyers; they may implode without warning.* Two versions are shown. I think the setup using the filter flask is more flexible. The screw clamp allows you to let air into your setup at a controlled rate. You might clamp the water trap to a ring stand when you use it. The connecting hoses have been known to flip unsecured flasks two out of three times.

WORKING WITH A MIXED-SOLVENT SYSTEM—THE GOOD PART

If, after sufficient agony, you cannot find a single solvent to recrystallize your product from, you may just give up and try a *mixed-solvent system.* Yes, it does mean you mix more than one solvent and *recrystallize using the mixture.* It should only be so easy. Sometimes you are told what the mixture is and the correct proportions. Then it is easy.

For an example, I could use "solvent 1" and "solvent 2," but that's clumsy. So I'll use the ethanol–water system and point out the interesting stuff as I go along.

The Ethanol—Water System

If you look up the solubility of water in ethanol (or ethanol in water), you find an ∞. This means they mix in all proportions. Any amount of one dissolves completely in the other—no matter what. Any volumes, any weights. You name it. The special word for this property is **miscibility.** Miscible solvent systems are the kinds you should use for mixed solvents. They keep you out of trouble. You'll be adding amounts of water to the ethanol, and ethanol to the water. If the two weren't miscible, they might begin to separate and form two layers as you changed the proportions. Then you'd have REAL trouble. So, go ahead. You *can* work with mixtures of solvents that aren't miscible. But don't say you haven't been warned.

The ethanol–water mixture is useful because:

1. *At high temperatures, it behaves like alcohol!*
2. *At low temperatures, it behaves like water!*

From this, you should get the idea that it would be good to use a mixed solvent to recrystallize compounds that are *soluble in alcohol* yet *insoluble in water.* You see, each solvent alone cannot be used. If the material is soluble in alcohol, not many crystals come back from alcohol alone. If the material is insoluble in water, you cannot even begin to dissolve it. So, you have a *mixed solvent,* with the best properties of *both* solvents. To actually perform a *mixed-solvent recrystallization,* you:

1. Dissolve the compound in the smallest amount of *hot ethanol.*
2. Add *hot water* until the solution turns cloudy. This **cloudiness** is *tiny crystals of compound coming out of solution.* Heat this solution to dissolve the crystals. If they do not dissolve completely, add *a very little hot ethanol* to force them back into solution.
3. Cool, and collect the crystals on a **Buchner funnel.**

Any solvent pair that behaves the same way can be used. The addition of hot solvents to one another can be tricky. It can be *extremely dangerous* if the boiling points of the solvents are very different. For the *water–methanol mixed solvent,* if 95°C water hits *hot methanol* (bp 65.0°C), watch out!

There are other miscible, mixed-solvent pairs—petroleum ether and diethyl ether, methanol and water, and ligroin and diethyl ether among them.

A MIXED-SOLVENT SYSTEM—THE BAD PART

Every silver lining has a cloud. More often than not, compounds "recrystallized" from a mixed-solvent system don't form crystals. Your compound may form an *oil* instead.

Oiling out is what it's called; more work is what it means. Compounds usually oil out if *the boiling point of the recrystallization solvent is higher than the melting point of the compound,* although that's not the only time. In any case, if the oil solidifies, the impurities are trapped in the now-solid "oil," and you'll have to purify the solid again.

Don't think you won't ever get oiling out if you stick to single, unmixed solvents. It's just that with two solvents, there's a greater chance you'll hit on a composition that will cause this.

Temporarily, you can:

1. Add more solvent. If it's a mixed-solvent system, try adding more of the solvent the solid is NOT soluble in. Or add more of the OTHER solvent. No contradiction. The point is to *change the composition.* Whether single solvent or mixed solvent, changing the composition is one way out of this mess.

2. Redissolve the oil by heating; then shake up the solution as it cools and begins to oil out. When these smaller droplets finally freeze out, they may form crystals that are relatively pure. They may not. You'll probably have to clean them up again. Just don't use the same recrystallization solvent.

Sometimes, once a solid oils out it doesn't want to solidify at all, and you might not have all day. Try removing a sample of the oil with an eyedropper or a disposable pipet. Then get a glass surface (watch glass) and add a few drops of a solvent that the compound is known to be *insoluble* in (usually water). Then use the rounded end of a glass rod to *triturate the oil with the solvent.* **Trituration** can be described loosely as beating an oil into a crystalline solid. Then you can put these crystals back into the rest of the oil. Possibly they'll act as seed crystals and get the rest of the oil to solidify. Again, you'll still have to clean up your compound.

SALTING OUT

Sometimes you'll have to recrystallize your organic compound from water. No big deal. But sometimes your organic compound is more than ever-so-slightly insoluble in water, and not all the compound will come back. Solution? Salt solution! A pinch of salt in the water raises the **ionic strength.** There are now charged ions in the water. Some of the water that solvated your compound goes with the salt ions. Your organic compound does not particularly like charged ions anyway, so more of your organic compound comes out of the solution.

You can dissolve about 36 g of common salt in 100 mL of cold water. That's the upper limit for salt. You can estimate how much salt you'll need to practically saturate the water with salt. Be careful, though—if you use too much salt, you may find yourself collecting salt crystals along with your product. (See also the application of salting out when you have to do an extraction, "Extraction Hints," in Chapter 15.)

FIGURE 13.9 Folding filter paper into eighths.

WORLD-FAMOUS FAN-FOLDED FLUTED PAPER

Some training in origami is *de rigueur* for chemists. It seems that the regular filter paper fold is inefficient, since very little of the paper is exposed. The idea here is to **flute** or **corrugate** the paper, increasing the surface area that is in contact with your filtrate. You'll have to do this several times to get good at it.

Right here let's review the difference between **fold** and **crease.** Folding is folding; creasing is folding, then stomping on it, and running fingers and fingernails over a fold over and over and over. Creasing so weakens the paper, especially near the folded point, that it may break at an inappropriate time in the filtration.

1. Fold the paper in half, then in half again, then in half again (Fig. 13.9). Press on this wedge of paper to get the fold lines to stay, but *don't crease. Do this in one direction only.* Always fold either toward you or away from you, but don't do both.

2. Unfold this cone *twice* so it looks like a semicircle (Fig. 13.10), and then put it down on a flat surface. Look at it and think for not less than two full minutes the first time you do this.

3. OK. Now try a "fan fold." You alternately fold, first in one direction and then the other, every individual eighth section of the semicircle (Fig. 13.11).

4. Open the fan and play with it until you get a fairly fluted filter cone (Fig. 13.12).

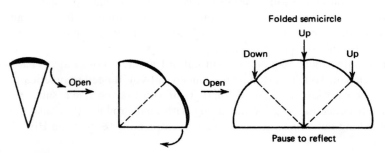

FIGURE 13.10 Unfolding to a sort of bent semicircle.

FIGURE 13.11 Refolding to a fan.

5. It'll be a bit difficult, but try to find the two opposing sections that are not folded correctly. Fold them inward (Fig. 13.12), and you'll have a fantastic fan-folded fluted filter paper of your very own.

P.S. For those with more money than patience, prefolded fan-folded fluted filter paper is available from suppliers.

EXERCISES

1. An impatient lab student plunges his hot, totally dissolved solution into his ice-water bath to get the crystals out quickly. Comment on the purity of his recrystallized product.

2. You perform the recrystallization procedure and get no crystals. What might you do to rectify this situation? Would it help to recall a Prego spaghetti sauce commercial that, talking about the composition of the sauce, stated, "It's in there!"?

3. OK, you have an oil, and you have to change the composition of your mixed-solvent system to get to a different point on the phase diagram. Do you add more of the more-polar solvent or the less-polar solvent?

FIGURE 13.12 Finishing the final fluted fan.

RECRYSTALLIZATION: Microscale

Microscale recrystallizations are done the same way the big ones are. You just use teeny-tiny glassware. Read the general section on crystallization first. Then come back here, and:

1. Put your solid in a small test tube.

2. Add a **microboiling stone** or a tiny **magnetic stirring vane** if you're on a stirring hot plate.

3. *Just cover* the solid with recrystallization solvent.

4. Heat this mix in the sand bath; stir with the spin bar or shake the test tube.

5. If the solid isn't dissolving, *consider:*
   a. The stuff not dissolving is insoluble trash; adding lots more solvent won't *ever* dissolve it.
   b. The stuff not dissolving is your compound; you need to add 1–2 more drops of solvent.
   This is a decision you'll have to make based on your own observations. Time to take a little responsibility for your actions. I wouldn't, however, add much more than 2–3 mL at the microscale level.

6. OK. Your product has *just dissolved* in the solvent. Add another 10% or so of excess solvent to keep the solid in solution during the hot filtration. This could be half a drop or so. Careful. If you're going to use *decolorizing carbon,* add it now; the whole tube should turn black. Heat for 5–10 seconds, and then use Pasteur pipets to filter. You might have to filter twice if you've used carbon.

7. Get your solution into a test tube to cool and recrystallize. Let it cool slowly to room temperature; then put the tube in an ice-water bath for final recrystallization.

8. "I didn't get any crystals." Heat the tube and boil off some of the solvent. (*Caution—Bumping!*) Use a microboiling stone, and don't point the tube at anyone. Repeat steps 7 and 8 until you get crystals. The point is to get to *saturation* while hot (all the solid that can be dissolved in the solvent), and the crystals will come out when cold.

If you've compared this outline to the large-scale recrystallization, you'll find that, with one exception—keeping a reservoir of hot solvent ready—the only difference here is the size of the equipment. Microscale means test tubes for flasks, filter pipets for funnels with paper, and so on. Because you can easily remove solvent at this scale, adding too much solvent is not quite the time-consuming boiling-off process you'd have to do on a larger scale.

ISOLATING THE CRYSTALS

First, see the section on "Pipet Filtering—Solids" in Chapter 7 where you remove solvent with a pipet. I also mention using a Hirsch funnel. Just reread the section on the Buchner funnel filtration (Chapter 13, "Recrystallization"), and where you see

"Buchner" substitute "Hirsch" and paste copies of the Hirsch funnel drawing over the Buchner funnel drawing. Just realize that the tiny disk of paper can fly up more easily. Think small.

CRAIG TUBE FILTRATION

Usually this is called Craig tube crystallization, because you've pipet-filtered your hot solution into the bottom of a Craig tube. So if you've recrystallized in something else, redissolve the crystals and get this solution into the bottom of a Craig tube (Fig. 14.1).

1. Carefully put the upper section of the Craig tube into the bottom section. Let the solution slowly cool to room temperature and then put the tube in an ice-water bath.

2. Now the fun part. You've got to turn this thing upside down, put it in a centrifuge tube, and *not* have it come apart and spill everything. Easy to say—"Solvent is now removed by inverting the Craig tube assembly into a centrifuge tube . . ." (Mayo *et al.*, p. 80). Easy to do? Well . . .

3. Get your centrifuge tube, and cut a length of copper wire as long as the tube. Make a loop in the end that's big enough to slip over the end of the Craig tube stem (Fig. 14.2). Twist the wire so that the loop won't open if you pull on it a bit.

Top of Craig tube

Ground glass joint

Bottom of Craig tube

Solvent

Crystals

FIGURE 14.1 Crystals in Craig tube.

Loop to fit end
of Craig tube

Close right-angle
bend

Wire as long as
the centrifuge tube
you'll be using

FIGURE 14.2   Preparing wire loop for Craig tube.

4. Bend the loop so that it makes about a right angle from the wire (Fig. 14.2). Ready?

5. Hold the bottom of the Craig tube between your thumb and forefinger. Put the loop over the end of the tube (Fig. 14.3).

6. Hold the wire gently between your thumb (or forefinger) and the glass tube bottom. Grip the hanging wire, and gently pull it downward so that the Craig tube top is pulled into the bottom (Fig. 14.3). Too much force here and you could snap the stem. Watch it.

7. Once this is adjusted, tightly press the wire to the glass, so that the wire will keep the Craig tube closed when you turn the tube upside down.

8. Put the centrifuge tube upside down over the top of the Craig tube (Fig. 14.4*b*). Watch it. If the fit is snug, the copper wire can cause problems, such as breaking the tubes. If the centrifuge tube is too tight, you might consider getting a wider one.

9. Once you get to this point, you may have three situations:
   Baby Bear (Fig. 14.4*a*): The centrifuge tube is a bit shorter than the Craig tube. Some of the Craig tube peeks out from the centrifuge tube. The centrifuge *might* have clearance in the head to accommodate the extra length. I don't like this. Perhaps you should ask your instructor. Invert the setup and get ready to centrifuge.

Wire loop down as far as possible

Wire kept close to Craig tube

Wire pinched between thumb and forefinger and Craig tube

FIGURE 14.3 Wiring the Craig tube.

Mama Bear (Fig. 14.4*b*): The centrifuge tube is a bit longer than the Craig tube. "A bit" here can vary. I mean you can push the Craig tube up into the centrifuge tube with your fingers so that the Craig tube stem touches the bottom or is within a centimeter or so of the bottom. Push the tube up as far as it will go. Invert the setup, and get ready to centrifuge.

Papa Bear (Fig. 14.4*c*): The centrifuge tube is a *lot* longer than the Craig tube. You can't push it up to the bottom with your fingers. Grab the copper wire at the end, invert the setup, and quickly *lower* the Craig tube to the bottom of the centrifuge tube. Now you're ready to centrifuge.

CENTRIFUGING THE CRAIG TUBE

Just a few things here:

1. "Where does the centrifuge go when it takes a walk?" The centrifuge must be *balanced*—an even number of tubes that have almost identical weights, set in positions opposite each other.

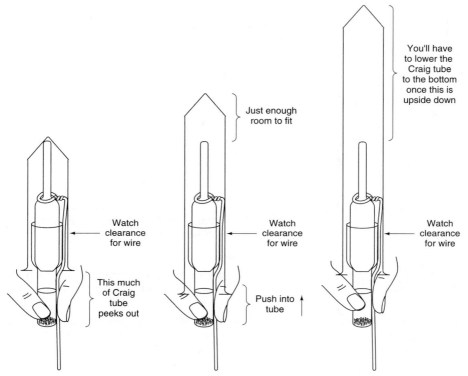

You'll have to lower the Craig tube to the bottom once this is upside down

Just enough room to fit

Watch clearance for wire

Watch clearance for wire

Watch clearance for wire

This much of Craig tube peeks out

Push into tube

FIGURE 14.4   The Three Bears of centrifuge tubes meet the Craig tube.

2. If you've cut the copper wire properly, it should stick out only a little bit from the centrifuge tube. You wouldn't want to be flailed by a foot-long copper wire, would you? If necessary, bend it out of the way. Don't bend it so that you lose the wire into the centrifuge tube and can't get it again.

3. If centrifuging 1 minute is good, centrifuging 15 minutes is *NOT* better. Whatever the directions call for, just do it. Your yield will not improve. Trust me.

Getting the Crystals Out

After you've centrifuged the Craig tube and the centrifuge has stopped turning,

Do not stop the centrifuge with your fingers! Patience!

Pull the centrifuge tubes out, and carefully lift out the Craig tube. The solvent will have filled the bottom of the centrifuge tube, and the crystals will be packed down near the Craig tube top (Fig. 14.5).

Use wire to pull
Craig tube out

Crystals
packed into
joint

Liquid at
bottom
of tube

FIGURE 14.5 Craig tube crystals after centrifugation.

I'd get a glass plate or watch glass, open the Craig tube, and put both sections on the glass to dry. (**CAUTION**—rolling Craig tubes will spread your product on the bench.)

After a bit, you can scrape your crystals into the appropriate tared container. (See "Tare to the Analytical Balance" in Chapter 5, "Microscale Jointware.") Don't try to get all of them; it's usually not worth the effort.

EXTRACTION AND WASHING

- *Add a few drops of a layer to 1 mL of water in a test tube: If it dissolves, it's an aqueous layer; if not, it's organic.*

- *Burp your funnels.*

- *Prevent vacuum buildup. Take the top stopper out when you empty a funnel.*

Extraction is one of the more complex operations that you'll do in the organic chemistry lab. For this reason, I'll go over it especially slowly and carefully. Another term that you'll see used simultaneously with **extraction** is **washing.** That's because extraction and washing are really the same operation, but each leads to a different end. How else to put this?

Let's make some soup. Put the vegetables, fresh from the store, in a pot. You run cold water in and over them to clean them and throw this water down the drain. Later, you run water in and over them to cook them. You keep this water—it's the soup.

Both operations are similar. Vegetables in a pot in contact with water the first time is a **wash.** You remove unwanted dirt. *You washed with water.* The second time, vegetables in a pot in contact with water is an **extraction.** You've **extracted** essences of the vegetables *into water.* Very similar operations; very different ends.

To put it a little differently,

> **You would extract good material from an impure matrix.**
> **You would wash the impurities from good material.**

The vegetable soup preparation is a *solid–liquid extraction.* So is coffee making. You extract some components of a solid directly into the solvent. You might do a solid–liquid extraction in lab as a separate experiment; *liquid–liquid extractions are routine.* They are so common that if you are told to do an extraction or a washing, *it is assumed* you will use *two liquids—two insoluble liquids—and a separatory funnel.* The separatory funnel, called a **sep funnel** by those in the know, is a special funnel that you can separate liquids in. You might look at the section on separatory funnels (later in this chapter) right now, and then come back later.

Two insoluble liquids in a separatory funnel will form **layers;** one liquid will float on top of the other. You usually have compounds dissolved in these layers, and either the compound you want is *extracted from one to the other,* or junk you don't want is *washed from one layer to the other.*

Making the soup, you have *no* difficulty deciding what to keep or what to throw away. First you throw the water away; later you keep it. But this can change. In a sep funnel, the layer that you want to keep one time may not be the layer that you want to keep the next time. Yet, if you throw one layer away prematurely, you are doomed.

NEVER-EVER LAND

> **Never, never, never, never,**
> **ever throw away any layer until you are absolutely sure you'll**
> **never need it again. Not very much of your product can be**
> **recovered from the sink trap!**

I'm using a word processor, so I can copy this warning over and over again, but let's not get carried away. One more time. Wake Up Out There!

Never, never, never, never,
ever throw away any layer until you are absolutely sure you'll
never need it again. Not very much of your product can be
recovered from the sink trap!

STARTING AN EXTRACTION

To do any extraction, you'll need two liquids or solutions. *They must be insoluble in each other.* **Insoluble** here has a practical definition:

When mixed together, the two liquids form two layers.

One liquid will float on top of the other. A good example is ether and water. Handbooks say that ether is slightly soluble in water. When ether and water are mixed, yes, some of the ether dissolves; most of the ether just floats on top of the water.

Really soluble or *miscible liquid pairs are no good* for extraction and washing. When you mix them, *they will not form two layers!* In fact, they'll *mix in all proportions.* A good example of this is acetone and water. What kinds of problems can this cause? Well, for one, *you cannot perform any extraction with two liquids that are miscible.*

Let's try it. A mixture of, say, some mineral acid (is HCl all right?) and an organic liquid, "compound A," needs to have that acid washed out of it. You dissolve the compound A–acid mixture in some acetone. It goes into the sep funnel, and now you add water to wash out the acid.

Acetone is miscible in water. No layers form! You lose!

Back to the lab bench. Empty the funnel. Start over. This time, having called yourself several colorful names because you should have read this section thoroughly in the first place, you dissolve the compound A–acid mixture in ether and put it into the sep funnel. Add water, and *two layers form!* Now you can wash the acid from the organic layer to the water layer. The water layer can be thrown away.

Note that the acid went into the water, *then the water was thrown out!* So we call this a **wash.** If the water layer had been saved, we'd say the acid had been **extracted into the water layer.** It may not make sense, but that's how it is.

Review:

1. You *must have two insoluble liquid layers* to perform an extraction.

2. *Solids must be dissolved in a solvent,* and that solvent must be insoluble in the other extracting or washing liquid.

3. If you are washing or extracting an organic liquid, dissolve it into another liquid, *just like a solid,* before extracting or washing it.

So these terms, **extraction** and **washing,** are related. Here are a few examples.

1. Extract with ether. Throw ether together with the solution of product, and pull out *only the product into the ether.*

2. Wash with 10% NaOH. Throw 10% NaOH together with the solution of product, and pull out *everything but product into the NaOH.*

3. You can even extract with 10% NaOH.

4. You can even wash with ether.

> *So extraction is pulling out what you want from all else! Washing is pulling out all else from what you want.*

And please note—*you always do the pulling from one layer into another.* That's also *two immiscible liquids.*

You'll actually have to do a few of these things before you get the hang of it, but bear with me. When your head stops hurting, reread this section.

DUTCH UNCLE ADVICE

Before I go on to the separatory funnel, I'd like to comment on a few questions I keep hearing when people do washings and extractions.

1. *"Which layer is the water layer?"* This is so simple, it confounds everyone. Get a small (10 × 75-mm) test tube and add about 0.5–1.0 mL water. Now add 2–4 drops of the layer you *think* is aqueous, and swirl the tube. If the stuff *doesn't dissolve* in the water, it's *not* an aqueous (water) layer. The stuff may sink to the bottom, float on the top, do *both,* or even *turn the water cloudy!* It will *not,* however, dissolve.

2. *"How come I got three layers?"* Sometimes, when you pour fresh water or some other solvent into the funnel, you get a small amount hanging at the top, and it looks like there are three different layers. Yes, it *looks* as if there are three different layers, but there are not three different layers, only two layers, where part of one has lost its way. Usually this mysterious third layer looks just like its parent, and you might gently swirl the funnel and its contents to reunite the family.

3. *"What's the density of sodium hydroxide?"* You've just done a wash with 5–10% sodium hydroxide solution, you've just read something about finding various layers in the funnel by their densities, and, by this question, you've just shown that you've missed the point. Most wash solutions are 5–10% active ingredient dissolved in water. This means they are 90–95% water. Looking up the density of the solid reagents, then, is a waste of time. The densities of these solutions are *very close* to that of water. (10% NaOH has a specific gravity of 1.1089.)

4. *"I've washed this organic compound six times with sodium bicarbonate solution, so why's it not basic yet?"* This involves finding the pH of the organic layer. I'll give it away right now: *You cannot find the pH of an organic layer.* Not directly. You find the pH of the aqueous layer that's been in contact with the organic layer. If the aqueous layer is on the top, dip a glass rod into it and touch the glass rod to your test paper. If the aqueous layer is on the bottom and your sep funnel is in a ring, let a drop of the aqueous layer out of the funnel to hang on the outlet tip. Transfer the drop to your test paper.

Do not ever dip the test paper into any solution. Ever.

Warning. Be *sure* you are testing the aqueous layer. Some organics are very tenacious and can get onto your glass rod. The organic layer may wet the test paper, but without water, any color you see doesn't mean much.

THE SEPARATORY FUNNEL

Before we go on to some practical examples, you might want to know more about where all this washing and extracting is carried out. I've mentioned that it's a special funnel called a **separatory funnel** (Fig. 15.1) and that you can impress your friends by calling it a **sep funnel.** Here are a few things you should know.

The Stopper

At the top of the sep funnel is a ⊺ glass stopper. There is a number, commonly ⊺ 22, possibly ⊺ 19/22, on the stopper head. *Make sure this number is on the head and that it is the same as the number marked on the funnel.* If this stopper is not so marked, you may find the product leaking over your shoes when you turn the sep funnel upside down. Try *not* to grease this stopper unless you plan to sauté your product. Unfortunately, these stoppers tend to get stuck in the funnel. The way out is to be sure you don't get the ground-glass surfaces wet with product. How? Pour solutions into the sep funnel as carefully as you might empty a shot glass of Scotch into the soda. Maybe *use a funnel.* To confuse matters, I'll suggest you use a light coating of grease. Unfortunately, my idea of light and your idea of light may be different.

Consult your instructor!

The Glass Stopcock

This is the time-honored favorite of separatory funnel makers everywhere. There is a notch at the small end that contains either a rubber ring *or* a metal clip, *but not both!* There are two purposes for the ring:

1. To keep the stopcock from falling out entirely. Unfortunately, the rubber rings are not aware of this, and the stopcock falls out anyway.

Plain stem

Ground-glass stem
with drip tip

FIGURE 15.1 Garden-variety separatory funnels.

2. To provide a sideways pressure, pulling the stopcock in, so that it will not leak. Names and addresses of individuals whose stopcocks could not possibly leak and did so anyway will be provided on request. So provide a little sideways pressure of your own.

When you grease a glass stopcock (and you must), do it very carefully so that the film of grease does *not* spread into the area of the stopper that comes in contact with any of your compound (Fig. 15.2).

The Teflon Stopcock

In wide use today, the **Teflon stopcock** (Fig. 15.3) requires no grease and will not freeze up! The glass surrounding the stopcock *is not ground glass and cannot be used in funnels that require ground-glass stopcocks!* The Teflon stopcocks

Clear, unbroken bands of grease

Rubber ring

Compression clip can substitute for rubber ring

FIGURE 15.2 The infamous glass stopcock.

are infinitely easier to take care of. A Teflon washer, a rubber ring, and, finally, a Teflon nut are placed on the threads of the stopcock. This nut holds the whole thing on. Any leakage at this stopcock results from:

1. A loose Teflon nut. Tighten it.
2. A missing Teflon washer or rubber ring. Replace it.
3. An attempt to place the wrong size or wrong taper Teflon stopcock into the funnel. This is extremely rare. Get a new funnel.

Loosen Teflon nut and pop the stopcock before storing

Pull gently

Teflon nut

Rubber ring

Teflon washer

FIGURE 15.3 Extreme close-up of Teflon stopcock-popping ritual.

Emergency stopcock warning!

Teflon may not stick, *but it sure can flow!* If the stopcock is extremely tight, the Teflon will bond itself to all the nooks and crannies in the glass in interesting ways. When you're through, always loosen the Teflon nut and "pop the stopcock" by pulling on the handle. The stopcock should be loose enough to spin freely when spun with one finger—*just remember to tighten it again before you use it.*

It seems to me that I'm the only one who reads the little plastic bags that hold the stopcock parts. Right on the bags it shows that after the stopcock goes in, *the Teflon washer goes on the stem first,* followed by the rubber ring, and then the Teflon nut (Fig. 15.3). So why do I find most of these things put together incorrectly?

HOW TO EXTRACT AND WASH WHAT

Here are some practical examples of washings and extractions, covering various types and mixtures and separations, broken down into the four classifications listed previously.

1. *A strong organic acid.* Extract into saturated (sat'd) sodium bicarbonate solution. **CAUTION!** Foaming and fizzing and spitting and all sorts of carrying on. The weak base turns the strong acid into a salt, and the salt dissolves in the water–bicarbonate solution. Because of all the fizzing, you'll have to be very careful. Pressure can build up and blow the stopper out of the funnel. Invert the funnel. *Point the stem away from everyone up and toward the back of the hood*—and open the stopcock to vent or "burp" the funnel.
   a. *To recover the acid,* add concentrated (conc'd) HCl until the solution is acidic. Use pH or litmus paper to make sure. Yes, the solution really fizzes and bubbles. You should use a large beaker, so that material isn't thrown onto the floor if there's too much foam.
   b. *To wash out the strong acid,* just throw the solution of bicarbonate away.
2. *A weakly acidic organic acid.* Extract into 10% NaOH–water solution. The strong base is needed to rip the protons off weak acids (they don't want to give them up) and turn them into salts. Then they'll go into the NaOH–water layer.
   a. *To recover the acid,* add conc'd HCl until the solution of base is acidic when tested with pH or litmus paper.
   b. *To wash out the weak acid,* just throw this NaOH–water solution away.
3. *An organic base.* Extract with 10% HCl–water solution. The strong acid turns the base into a salt. (This *turning the whatever into a salt that dissolves in the water solution* should be pretty familiar to you by now. Think about it.) Then the salt goes into the water layer.
   a. *To recover the base,* add ammonium hydroxide to the water solution until the solution is *basic* to pH or litmus paper. *Note that this is the reverse of the treatment given to organic acids.*

b. *To wash out an organic base, or any base,* wash as previously noted and throw out the solution.

4. *A **neutral organic**.* If you've extracted strong acids first, then weak acids, and then bases, there are only neutral compound(s) left. If possible, just remove the solvent that now contains *only your neutral compound.* If you have *more than one neutral compound,* you may want to extract one from the other(s). You'll have to find *two different immiscible organic liquids,* and *one liquid must dissolve only the neutral organic compound you want!* A tall order. You must count on *one neutral organic compound being more soluble in one layer than in the other.* Usually the separation is *not clean—not complete.* And you have to do more work.

What's "more work"? That depends on the results of your extraction.

The Road to Recovery—Back-Extraction

I've mentioned **recovery** of the four types of extractable materials, but that's not all the work you'll have to do to get the compounds in shape for further use.

1. If the recovered material is *soluble in the aqueous recovery solution,* you'll have to do a **back-extraction.**
   a. Find a solvent that dissolves your compound and is *not miscible in the aqueous recovery solution.* This solvent should boil at a low temperature ($<100°C$), since you will have to remove it. Ethyl ether is a common choice. (**Hazard!** *Very flammable.*)
   b. *Then you extract your compound back from the aqueous recovery solution into this organic solvent.*
   c. Dry this organic solution with a drying agent (see Chapter 10, "Drying Agents").
   d. Now you can remove the organic solvent. Either distill the mixture or evaporate it, perhaps on a steam bath. All this is done away from flames and in a hood.
   When you're through removing the solvent and your product is not pure, clean it up. If your product is a liquid, you might distill it; if it is a solid, you might recrystallize it. Make sure it is clean.

2. If the recovered material is *insoluble in the aqueous recovery solution* and *it is a solid,* collect the crystals on a Buchner funnel. If they are *not pure,* you should recrystallize them.

3. If the recovered material is *insoluble in the aqueous recovery solution* and *it is a liquid,* you can use your separatory funnel to *separate the aqueous recovery solution from your liquid product. Then dry your liquid product and distill it if it is not clean. Or you might just do a back-extraction* as just described. This has the added advantage of removing the small amount of liquid product that dissolves in the aqueous recovery solution and increases your yield. Remember to dry the back-extracted solution before you remove the organic solvent. Then distill your liquid compound if it is not clean.

A Sample Extraction

I think the only way I can bring this out is to use a typical example. This may ruin a few lab quizzes, but if it helps, it helps.

Say you have to separate a mixture of *benzoic acid (1), phenol (2), p-toluidine (4-methylaniline) (3),* and *anisole (methoxybenzene) (4)* by extraction. The numbers refer to the class of compound, as previously listed. We're assuming that none of the compounds reacts with any of the others and that you know we're using all four types as indicated. Phenol and 4-methylaniline are corrosive toxic poisons, and if you get near these compounds in lab, *be very careful.* When they are used as an example on these pages, however, you are quite safe. Here's a sequence of tactics.

1. Dissolve the mixture in ether. Ether is insoluble in the water solutions you will extract into. Ether happens to dissolve all four compounds. Aren't you lucky? You bet! It takes lots of hard work to come up with the "typical student example."

2. Extract the ether solution with 10% HCl. This converts *only* compound 3, the basic *p*-toluidine, into the hydrochloride salt, which dissolves in the 10% HCl layer. You have just *extracted a base with an acid solution.* Save this solution for later.

3. Now extract the ether solution with sat'd sodium bicarbonate solution. *Careful!* Boy, will this fizz! Remember to swirl the contents and release the pressure. The weak base converts *only* compound 1, the *benzoic acid, to a salt, which dissolves in the sat'd bicarbonate solution.* Save this for later.

4. Now extract the ether solution with the 10% NaOH solution. This converts compound 2, *weak acid,* phenol, to a salt, which dissolves in the 10% NaOH layer. Save this for later. *If you do this step before step 3, that is, extract with 10% NaOH solution before the sodium bicarbonate solution, both the weak acid,* phenol, *and the strong acid,* benzoic acid, *will be pulled out into the sodium hydroxide.* Ha-ha. This is the usual kicker they put in lab quizzes, and people always forget it.

5. The only thing left is the neutral organic compound dissolved in the ether. Just drain this solution into a flask.

So, now we have four flasks with four solutions with one component in each. *They are separated.* You may ask, "How do we get these back?"

1. *The basic compound (3).* Add ammonium hydroxide until the solution turns basic (test with litmus or pH paper). The *p*-toluidine, or organic base (3), is regenerated.

2. *The strong acid or the weak acid (1, 2).* A bonus. Add dilute HCl until the solution turns acidic to an indicator paper. Do it to the other solution. Both acids will be regenerated.

3. *The neutral compound (4).* It's in the ether. If you evaporate the ether (*no flames!*), the compound should come back.

Now, when you recover these compounds, sometimes they don't come back in such good shape. You will have to do more work.

1. Addition of HCl to the benzoic acid extract will produce huge amounts of white crystals. Get out the Buchner funnel and have a field day! Collect all you want. But they won't be in the best of shape. Recrystallize them. (**Note:** *This compound is insoluble in the aqueous recovery solution.*)

2. The phenol extract is a different matter. You see, *phenol is soluble in water,* and it doesn't come back well at all. So, get some fresh ether, extract the phenol from HCl solution to the ether, and evaporate the ether. Sounds crazy, no? No. Remember, I called this a **back-extraction,** and you'll have to do this more often than you would like to believe. (**Note:** *This compound is soluble in the aqueous recovery solution.*)

3. The *p*-toluidine should return after the addition of ammonium hydroxide. Recrystallize it from ethanol, so that it also looks respectable again.

4. The neutral anisole happens to be a *liquid (bp 155°C),* and you'll have to take care when you evaporate the ether that you don't lose much anisole. Of course, you shouldn't expect to see any crystals. Now this neutral anisole liquid that comes back after you've evaporated the ether (*no flames!*) will probably be contaminated with a little bit of all of the other compounds that started out in the ether. You will have to purify this liquid, probably by a simple distillation.

You may or may not have to do all of this with the other solutions or with any other solution that you ever extract in your life. You must choose. Art over science. As confusing as this is, I have simplified it a lot. Usually you have to extract these solutions more than once, and the separation is not as clean as you'd like. Not 100% but pretty good. If you are still confused, see your instructor.

Performing an Extraction or Washing

1. Suspend a sep funnel in an iron ring.

2. Remove the stopper.

3. *Make sure the stopcock is closed!* You don't really want to scrape your product off the bench top.

4. Add the solution to be extracted or washed. Less than half full, please. Add the extraction or washing solvent. An equal volume is usually enough. The funnel is funnel shaped, and the equal volumes won't look equal.

5. Replace the stopper.

6. Remove the sep funnel from the iron ring. Hold the stopper and stopcock tightly. Pressure may build up during the next step and blow your product out onto the floor.

Escaping gases

Danger!
Point the stem of the funnel
<u>away from everyone!</u>
(including yourself!)

Hold <u>both</u> the stopper
and the stopcock very tightly!!

Liquid phases

Hold here <u>tightly!</u>

FIGURE 15.4 Holding a
sep funnel so as not to get
stuff all over.

7. Invert the sep funnel (Fig. 15.4).

Point the stem up away from everyone—
up into the back of a hood if at all possible!

Make *sure* the liquid has drained down away from the stopcock, then *slowly*
open the stopcock. You may hear a whoosh, possibly a pffffft, as the pressure
is released. This is due to the high vapor pressure of some solvents or to a gas
evolved from a reaction during the mixing. This can cause big trouble when
you are told to neutralize acid by washing with sodium carbonate or sodium
bicarbonate solutions.

8. Close the stopcock!

9. Shake the funnel gently, invert it, and open the stopcock again.

10. Repeat steps 8 and 9 until no more gas escapes.

11. If you see that you might get an **emulsion**—*a fog of particles*—with this gentle
inversion, *do NOT shake the funnel vigorously.* You might have to continue the
rocking and inverting motions 30 to 100 times, as needed, to get a separation.
Check with your instructor and with the hints on breaking up emulsions (see
"Extraction Hints," following). Otherwise, shake the funnel vigorously about
10 times to get good distribution of the solvents and solutes. Really shake it.

12. Put the sep funnel back in the iron ring.

13. *Remove the glass stopper.* Otherwise the funnel won't drain, and you'll waste your time just standing there.

14. Open the stopcock, and let the bottom layer drain off into a flask.

15. Close the stopcock, swirl the funnel gently, and then wait to see if any more of the bottom layer forms. If so, collect it. If not, assume you got it all in the flask.

16. Let the remaining layer out into another flask.

To extract any layer again, return that layer to the sep funnel, *add fresh extraction or washing solvent,* and repeat this procedure starting from step 5.

Never, never, never, never,
ever throw away any layer until you are absolutely sure you'll
never need it again. Not very much of your product can be
recovered from the sink trap!

Extraction Hints

1. Several smaller washings or extractions are better than one big one (see Chapter 35, "Theory of Extraction").

2. Extracting or washing a layer *twice,* perhaps three times, is usually enough. Diminishing returns set in after that.

3. Sometimes you'll have to find out which layer is **the water layer.** This is so simple, it confounds everyone. Add 2–4 drops of each layer to a test tube containing 1 mL of water. Shake the tube. If the stuff *doesn't dissolve* in the water, it's *not* an aqueous (water) layer. The stuff may sink to the bottom, float on the top, do *both,* or even *turn the water cloudy!* It will *not,* however, dissolve.

4. If *only the top layer* is being extracted or washed, *it does not have to be removed from the funnel,* ever. Just drain off the bottom layer, and then add more *fresh* extraction or washing solvent. Ask your instructor about this.

5. You *can* combine the extracts of a multiple extraction, *if they have the same material in them.*

6. If you have to wash your organic compound with water and the organic is *slightly soluble in water,* try washing with *saturated salt solution.* The theory is that if all that salt dissolved in the water, what room is there for your organic product? This point is a favorite of quizmakers and should be remembered. It's the same thing that happens when you add salt to reduce the solubility of your compound during a crystallization (see "Salting Out" in Chapter 13).

7. If you get an **emulsion,** you have not two distinct layers, but a kind of a *fog of particles.* Sometimes you can break up the charge on the suspended droplets by adding a little salt or some acid or base. Or add ethanol. Or stir the solutions

slowly with a glass rod. Or gravity-filter the entire contents of your separatory funnel through filter paper. Or laugh. Or cry. Emulsion-breaking is a bit of an art. Careful with the acids and bases, though. They can react with your product and destroy it.

8. If you decide to add salt to a sep funnel, don't add so much that it clogs up the stopcock! For the same reason, keep drying agents out of sep funnels.

9. Sometimes some material comes out, or it will not dissolve in the two liquid layers and hangs in there at the **interface.** It may be that there's not enough liquid to dissolve this material. One cure is to *add more fresh solvent of one layer or the other.* The solid may dissolve. If there's no room to add more, you may have to remove *both* (yes, both) layers from the funnel, and try to dissolve this solid in either of the solvents. It can be confusing. If the material does *not* redissolve, then it is a new compound and should be saved for analysis. You should see your instructor for that one.

EXERCISES

1. What are the characteristics of a good extraction solvent?

2. In a typical "Extraction of Caffeine" experiment, you use boiling water to get the caffeine out of the coffee or tea, and then extract this aqueous solution with diethyl ether or methylene chloride. Why can't you immediately put the aqueous solution into the sep funnel with the ether or methylene chloride in order to save time?

3. Compare and contrast extraction with washing.

4. The Teflon stopcock of separatory funnels is kept on the funnel by a Teflon nut, a Teflon washer, and a rubber ring. What is the correct order, starting from the glass? Do your separatory funnels have these things on in the correct order?

EXTRACTION AND WASHING: Microscale

- *Guard against vapor pressure buildup in pipets; your product might blow out.*

In microscale, instead of a sep funnel you use a conical vial and some Pasteur pipets. First you mix your extraction solvent with your product; then you separate the two liquids.

MIXING

1. Put the material to be extracted into an appropriately sized conical vial. This vial should be, at *minimum,* twice the volume of the liquid you want to extract. Usually, this conical vial is the reaction vial for the experiment, so the choice is easily made (forced on you).

2. Add an appropriate solvent (*not*—repeat—*not* extracting liquid), so the volume of the mixture is about 1 mL. (What if it's about 1 mL already?)

3. Add about 1 mL of extracting solvent to the vial.

4. At this point, cap the vial and shake to mix (**Caution**—leaks!). Better you should add a **magnetic spinning vane** and spin the two layers for a minute or so.

SEPARATION: REMOVING
THE BOTTOM LAYER (FIG. 16.1)

1. Have an empty vial ready.

2. Prewet a Pasteur filter pipet and bulb combo with the extraction solvent. (See Chapter 7, "Pipet Tips.")

Pasteur filter pipet

CH_2Cl_2

H_2O

CH_2Cl_2

STEP I

STEP II

STEP III

FIGURE 16.1 Pipet removes bottom layer.

3. Squeeze the bulb.

4. Stick the pipet into the vial down to the bottom.

5. Slowly—read that word again—slowly, draw the bottom layer up into the pipet. Get it all now. (Actually, you could do this a few times if necessary.) And don't suck any liquid *into* the rubber bulb!

6. Squirt the bottom layer into an empty vial.

SEPARATION: REMOVING THE TOP LAYER (FIG. 16.2)

1. Have an empty vial ready.

2. Prewet a Pasteur filter pipet and bulb combo with the extraction solvent. (See Chapter 7, "Pipet Tips.")

3. Really squeeze the bulb.

4. Stick the pipet into the vial down to the bottom.

5. Slowly—read that word again—slowly, draw *all*—yes *all,* I mean both layers—into the pipet. *Don't slurp!* If you draw lots of air into the pipet *after* both layers are in there, the air bubble will mix both layers in the pipet. Even worse, because of evaporation into this new "fresh" air, vapor pressure buildup can make both layers fly out back into the vial, onto the bench top, on to your hands, wherever. And don't suck any of this *into* the rubber bulb!

Ether

H₂O

Ether

Ether

H₂O

H₂O

STEP I   STEP II   STEP III

FIGURE 16.2 Pipet removes top layer.

6. Squeeze the bulb to put the *bottom* layer back into the original vial.

7. Squirt the top layer into an empty vial.

SEPARATION: REMOVING *BOTH* LAYERS

Some say, if you already have both layers in the pipet (Fig. 16.2), why bother squirting the lower layer, or any layer really, back into the vial? Suck 'em both up, and squirt each into whatever vessel you need to. "Squirt" slowly, gently. Remember, you're separating two 1-mL portions of liquid here. And it's probably better to leave a little of the organic layer in with the water layer here. Drying 1 mL of a liquid is going to be tricky enough without extra water from the water layer in there.

SOURCES OF HEAT

- *Burners are the last resort.*

Many times you'll have to heat something. Don't just reach for the Bunsen burner. That flame you start just may be your own. There are alternative sources you should think of *first*.

BOILING STONES

All you want to do is start a distillation. Instructor walks up and says,

"Use a boiling stone or it'll bump."

"But I'm only gonna . . ."

"Use a boiling stone or it'll bump."

"It's started already and . . ."

"Use a boiling stone or it'll bump."

"I'm not gonna go and . . ."

Suddenly—whoosh! Product all over the bench! Next time, you put a boiling stone in *before* you start. No bumping. But your instructor won't let you forget the time you did it your way.

Don't let this happen to you. Use a **brand-new boiling stone** every time you have to boil a liquid. A close-up comparison between a boiling stone and the inner walls of a typical glass vessel reveals thousands of tiny nucleating points on the stone where vaporization can take place, in contrast to the smooth glass surface that can hide unsightly hot spots and lead to bumping, a massive instantaneous vaporization that will throw your product all over.

CAUTION! Introducing a boiling stone into hot liquid may result in instant vaporization and loss of product. Remove the heat source, swirl the liquid to remove hot spots, and then add the boiling stone.

Used as directed, the boiling stone will relieve minor hot spots and prevent loss of product through bumping. So remember . . . whenever you boil, wherever *you boil,*

Always use a fresh boiling stone!

Don't be the last on your bench to get this miracle of modern science made exclusively from nature's most common elements.

THE STEAM BATH

If one of the components boils below 70°C and you use a *Bunsen burner,* you may have a hard time putting out the fire. Use a **steam bath** (Fig. 17.1)!

1. Find a steam tap. It's like a water tap, only this one dispenses steam. **CAUTION!** You can get burned.

2. Connect tubing to the tap *now.* It's going to get awfully hot in use. Make sure you've connected a piece that'll be long enough to reach your steam bath.

Clamp

Flask down in bath

Steam in

Water out to sink

FIGURE 17.1 The steam bath in use.

3. Don't connect this tube to the steam bath yet! Just put it into a sink. Because steam lines are usually full of water from condensed steam, *drain the lines first;* otherwise, you'll waterlog your steam bath.

4. **CAUTION!** Slowly open the steam tap. You'll probably hear bonking and clanging as steam enters the line. Water will come out. It'll get hotter and may start to spit.

5. Wait until the line is mostly clear of water, then turn off the steam tap. *Wait for the tubing to cool.*

6. Slowly, carefully, and cautiously, *making sure the tube is not hot,* connect the tube to the inlet of the steam bath. This is the *uppermost* connection on the steam bath.

7. Connect another tube to the outlet of the steam bath—the *lower* connection— and to a drain. Any water that condenses in the bath while you're using it will drain out. Make sure all parts of the drain hose are *lower* than the bath to keep the water flowing out of the bath.

Usually, steam baths have concentric rings as covers (Fig. 17.1). You can control the "size" of the bath by removing various rings.

Never do this after you've started the steam. You will get burned!

And don't forget—round-bottom flasks should be about halfway in the bath. You shouldn't really let steam rise up all around the flask. Lots of steam will certainly steam up the lab and may expose you to corrosion inhibitors (morpholine) in the steam lines.

THE BUNSEN BURNER

The first time you get the urge to take out a Bunsen burner and light it up, *don't.* You may blow yourself up. Please check with your instructor to see if you even need a burner. Once you find out that you *can* use a burner, assume that the person who used it last didn't know much about burners. Take some precautions so as not to burn your eyebrows off.

Now, Bunsen burners are not the only kind. There are **Tirrill burners** and **Meker burners** as well. Some are fancier than others, but they work pretty much the same. So when I say **burner** anywhere in the text, it could be any of them.

1. Find the round knob that controls the **needle valve.** This is at the base of the burner. Turn it fully clockwise (inward) to stop the flow of gas completely. If your burner doesn't have a needle valve, it's a traditional Bunsen burner, and the gas flow has to be regulated at the bench stopcock (Fig. 17.2). This can be dangerous, especially if you have to reach over your apparatus and burner to turn off the gas. Try to get a different model.

2. There is a **movable collar** at the base of the burner which controls air flow. For now, see that *all* the holes are closed (i.e., no air gets in).

3. Connect the burner to the bench stopcock by some tubing and turn the bench valve *full on.* The bench valve handle should be parallel to the outlet (Fig. 17.2).

4. Now, *slowly* open the needle valve. You may be just able to hear some gas escaping. Light the burner. *Mind your face!* Don't look down at the burner as you open the valve.

5. You'll get a wavy yellow flame, something you don't really want. But at least it'll light. Now open the air collar a little. The yellow disappears; a blue flame forms. This is what you want.

"Roaring" blue flame

Handle in same plane as outlet – full on

Handle at right angles for off

GAS

Sharp blue inner cone

Spin barrel or collar to regulate air flow

Air in here

Air intakes

Needle valve to regulate gas (not in the traditional Bunsen burner)

Typical bench gas valve, turned ON full blast

FIGURE 17.2   More than you care to know about burners.

6. Now, adjust the needle valve and collar (the adjustments play off each other) for a steady blue flame.

Burner Hints

1. Air does not burn. You must wait until the gas has pushed the air out of the connecting tubing. Otherwise, you might conclude that none of the burners in the lab work. Patience, please.

2. When you set up for distillation or reflux, don't waste a lot of time *raising* and *lowering* the entire setup so that the burner will fit. This is nonsense. Move the burner! Tilt it (Fig. 17.3). If you leave the burner motionless under the flask, you may scorch the compound, and your precious product can become a "dark intractable material."

3. Placing a wire gauze between the flame and the flask spreads out the heat evenly. Even so, the burner may have to be moved around. Hot spots can cause star cracks to appear in the flask (see "Round-Bottom Flasks" in Chapter 4, "Jointware").

4. *Never* place the flask in the ring without a screen (Fig. 17.4). The iron ring heats up faster than the flask, and the flask cracks in the nicest line around it you've ever seen. The bottom falls off, and the material is all over your shoes.

Screen

Iron ring

Hand-moved burner

FIGURE 17.3   Don't raise the flask, lower the burner.

FIGURE 17.4 Flask in the iron ring.

THE HEATING MANTLE

A very nice source of heat, the heating mantle takes some special equipment and finesse (Fig. 17.5).

1. *Variable-voltage transformer.* The transformer takes the quite lethal 120 V from the wall socket and can change it to an equally dangerous 0 to 120 V,

FIGURE 17.5 Round-bottom flask and mantle ready to go.

depending on the setting on the dial. Unlike temperature settings on a Mel-Temp, on a transformer 0 means 0 V, 20 means 20 V, and so on. I like to start at 0 V and work my way up. Depending on how much heat you want, values from 40 to 70 seem to be good places to start. You'll also need a cord that can plug into both the transformer and the heating mantle. (See Chapter 12, "The Melting Point Experiment.")

2. ***Traditional fiberglass heating mantle.*** An electric heater wrapped in fiberglass insulation and cloth that looks vaguely like a catcher's mitt (Fig. 17.5).

3. ***Thermowell heating mantle.*** You can think of the Thermowell heating mantle as the fiberglass heating mantle in a can. In addition, there is a hard ceramic shell that your flask fits into (Fig. 17.6). Besides just being more mechanically sound, it'll help stop corrosive liquids from damaging the heating element if your flask cracks while you're heating it.

4. ***Things not to do.***
   a. *Don't ever plug the mantle directly into the wall socket!* I know, the curved prongs on the mantle connection won't fit, but the straight prongs on the adapter cord will. Always use a variable-voltage transformer and start with the transformer set to zero.
   b. *Don't use too small a mantle.* The only cure for this is to *get one that fits properly.* The poor contact between the mantle and the glass doesn't transfer heat readily, and the mantle burns out.
   c. *Don't use too large a mantle.* The only good cure for this is to *get one that fits properly.* An acceptable fix is to *fill the mantle with sand after the flask is in* but before you turn the voltage on. Otherwise, the mantle will burn out.

Hint: When you set up a heating mantle to heat any flask, usually for **distillation** or **reflux,** put the mantle on an iron ring and keep it clamped a few inches above the bench top (Fig. 17.5). Then clamp the flask *at the neck* in case you have to remove the heat quickly. You can just unscrew the lower clamp and drop the mantle and iron ring.

FIGURE 17.6 A Thermowell heating mantle.

PROPORTIONAL HEATERS
AND STEPLESS CONTROLLERS

In all of these cases of heating liquids for distillation or reflux, we really control the *electric power* directly, not the heat or temperature. Power is applied to the heating elements, and they warm up. Yet the final temperature is determined by the heat loss to the room, the air, and, most important, the flask you're heating. There are several types of electric power controls.

1. ***The variable-voltage transformer.*** We've discussed this just previously. Let me briefly restate the case: Set the transformer to 50 on the 0–100 dial and you get 50% of the line voltage, all the time, night and day, rain or shine.

2. ***The mechanical stepless controller.*** This appears to be *the* inexpensive replacement for the variable-voltage transformer. Inside one model there's a small heating wire wound around a bimetal strip with a magnet at one end (Fig. 17.7). A plunger connected to the dial on the front panel changes the distance between the magnet and a metal plate. When you turn the device on with a heating mantle attached, current goes through the small heating wire and the mantle. The mantle is now on *full blast* (120 V out of 120 V from the electric wall socket)! As the small heating wire warms the bimetal strip, the strip expands, distorts, and finally pulls the magnet from its metal plate, opening the circuit. The mantle now cools rapidly (0 V out of 120 V from the wall socket), along with the bimetal strip.

FIGURE 17.7   Inside a mechanical stepless heater.

FIGURE 17.8 Light dimmer and heating mantle triac power control.

Eventually, the strip cools enough to let the magnet get close to that metal plate, and—CLICK—everything's on full tilt again.

The front panel control varies the **duty cycle,** the time the controller is *full on* to the time the controller is *full off.* If the flask, contents, and heating mantle are substantial, it takes a long time for them to warm up and cool down. A setup like that would have a large **thermal lag.** With small setups (approximately 50 mL or so), there is a small thermal lag, and very wild temperature fluctuations can occur. Also, operating a heating mantle this way is just like repeatedly plugging and unplugging it directly into the wall socket. Not many devices easily take that kind of treatment.

3. *The electronic stepless controller.* Would you believe a *light dimmer*? The electronic controller has a *triac,* a semiconductor device that lets fractions of the AC power cycle through to the heating mantle. The AC power varies like a sine wave, from 0 to 120 V from one peak to the next. At a setting of 25%, the triac remains off during much of the AC cycle, finally turning on when the time is right (Fig. 17.8). Although the triac does turn "full off and full on," it does so at times in so carefully controlled a way that the mantle never sees full line power (unless you deliberately set it there).

EXERCISE

Why shouldn't you plug a heating mantle directly into the electrical outlet on the bench?

CLAMPS AND CLAMPING

- *Plastic clips melt. Watch where you use them.*
- *Clamps move up a bit when tightened to the rod. Use this fact to make a tighter setup.*

Unfortunately, glass apparatus needs to be held in place with more than just spit and baling wire. In fact, you would do well to use clamps. Life would be simple if there were just one type of fastener, but that's not the case.

1. **The simple buret clamp** (Fig. 18.1). Though it's popular in other chem labs, the simple buret clamp just doesn't cut it for organic lab. The clamp is too short, and adjusting angles with the locknut (by loosening the locknut, swiveling the clamp jaws to the correct angle, and tightening the locknut against the *back* stop, away from the jaws) is not a great deal of fun. If you're not careful, the jaws will slip right around and all the chemicals in your flask will fall out.

2. **The simple extension clamp and clamp fastener** (Fig. 18.2). This two-piece beast is the second-best clamp going. It is much longer (approximately 12 in.), so you can easily get to complex setups. By loosening the **clamp holder thumbscrew,** the clamp can be pulled out, pushed back, or rotated to any angle. By loosening the **ring-stand thumbscrew,** the clamp, along with the clamp holder, can move up and down.

3. **The three-fingered extension clamp** (Fig. 18.3). This is truly the Cadillac of clamps, with a price to match. They usually try to confuse you with *two thumbscrews* to tighten, unlike the regular extension clamp. This gives a bit more flexibility, at the cost of a slightly more complicated way of setting up. You can make life simple by opening the *two-prong bottom jaw to a 10° to 20° angle from the horizontal and treating that jaw as fixed.* This will save a lot of wear and tear when you set equipment up, but you can *always move the bottom jaw* if you have to.

FIGURE 18.1 The "barely adequate for organic lab" buret clamp.

FIGURE 18.2 The extension clamp and clamp fastener.

FIGURE 18.3 The three-fingered clamp with clamp fastener.

CLAMPING A DISTILLATION SETUP

You'll have to clamp many things in your life as a chemist, and one of the more frustrating setups to clamp is the **simple distillation** (see Chapter 19, "Distillation"). If you can set this up, you probably will be able to clamp other common setups without much trouble. Here's one way to go about setting up the simple distillation.

1. OK, get a **ring stand,** an **extension clamp,** and a **clamp fastener,** and put them all together. What heat source? A Bunsen burner, and you'll need more room than you do with a heating mantle (see Chapter 17, "Sources of Heat"). In any case, you don't know where the receiving flask will show up, and then you might have to readjust the entire setup. Yes, you should have read the experiment before hand, so you'd know about the heating mantles.

2. Clamp the flask (around the neck) a few inches up the ring stand (Fig. 18.4). We *are* using heating mantles, and you'll need the room underneath to drop the

Three-fingered clamp

Heating mantle

Extension ring

Ring stand

FIGURE 18.4 Flask and heating mantle out on a ring stand.

mantle in case it gets too hot. That's why the flask is *clamped at the neck. Yes. That's where the flask is ALWAYS clamped, no matter what the heat source,* so it doesn't fall when the mantle comes down. What holds the mantle? Extension ring and clamp fastener.

3. Remember, whether you set these up from left to right, or right to left—*distilling flask first!*

4. Add the **three-way adapter** now (Fig. 18.5). Thermometer and thermometer adapter come later.

5. Put a second ring stand about one condenser length away from the first ring stand. Now add the condenser, just holding it onto the three-way adapter. Make a mental note of where the inner joint (end) of the condenser is (Fig. 18.6)— you'll want to put a clamp about there. Remove the condenser.

6. Get an extension clamp and clamp fastener. Open the jaws of the clamp so they're *wider* than the outer joint on a vacuum adapter. Set this clamp so that

FIGURE 18.5 Clamps and flask and three-way adapter.

Move condenser up
onto three-way
adapter

Note this position—
"remember it" for a
vacuum adapter clamp

Second ring stand

FIGURE 18.6   Trial fit for vacuum adapter.

it can accept the outer joint of a vacuum adapter at the angle and height you
made a mental note of in step 5.

7. Put a vacuum adapter on the end of your condenser, and hold it there. Put
   the top of the condenser (outer joint) onto the three-way adapter, and get
   the vacuum adapter joint cradled in the clamp you've set for it. Push the
   condenser/vacuum adapter toward the three-way adapter; lightly tighten the
   clamp on the vacuum adapter. The clamp *must* stop the vacuum adapter from
   slipping off the condenser (Fig. 18.7).

8. At this point, you have two clamps, one holding the distilling flask and the
   other at the end of the condenser/vacuum adapter (Fig. 18.8).

9. There are two ways to go now:

FIGURE 18.7 Correctly clamping the vacuum adapter.

a. The receiving flask will not fit vertically from the end of the vacuum adapter to the bench top (Fig. 18.8). Easy. With the vacuum adapter clamp *relaxed* (not so loose that the adapter falls off!), rotate the adapter toward you. There's no reason it *has* to be at a right angle. Stick a receiving flask on the end, put a suitable cork ring under the flask, and that's it. Note that you can *easily* change receiving flasks. Just slide one off and slide another on.

b. There's a lot of room from the end of the vacuum adapter to the bench top. In this case, you'll have to set up another clamp (possibly another ring stand) and clamp the flask at the neck (Fig. 18.9). This will make changing the receiving flask a bit more difficult; you'll have to unclamp one flask before slipping another in and reclamping the new one.

FIGURE 18.8   Distillation setup with receiving flask rotated toward you.

10. All the clamps are set up, all the joints are tight—now where is that thermometer adapter?

CLIPPING A DISTILLATION SETUP

There is another way of constructing a distillation setup: **Keck clips** (Fig. 18.10). These plastic clips have a large semicircle that gloms onto the outer joint, a smaller semicircle that hugs the inner joint, and three arms that hold the two semicircular clips and, as a consequence, hold the two parts of the glass joint together (Fig. 18.11). It's actually a lot of fun connecting jointware and clipping each joint together, watching the setup grow larger and larger as each new piece is clipped to the previous one.

As neat as Keck clips are, there is one small problem—they melt. If you clip together a distillation setup, and the compound in the flask boils at a relatively high temperature, the clip holding the distilling flask to the three-way adapter melts and then flows over the joint like melted cheese.

Never, ever use a plastic clip to attach a flask to a three-way adapter.

The clip that holds the three-way adapter to the condenser usually survives this, and the clips that hold the condenser to the vacuum adapter and the vacuum adapter to the receiving flask are not in any trouble at all.

FIGURE 18.9 Distillation setup with receiving flask at usual right-angle position.

FIGURE 18.10 Top view and side view of a Keck clip.

FIGURE 18.11 Hold on! It's the Keck clip in action.

The real fun starts when, after the setup cools and the plastic solidifies, you have to forcefully, yet carefully, chisel the globs of plastic off the joints.

Of course, this won't ever happen to you, oh reader of this manual, because you will use a clamp to hold the distilling flask, not a Keck clip. You know you may have to quickly lower the distilling flask, and if it's clipped, there's no way you can do that.

- *Gentle water flow through condensers—not Niagara Falls.*

- *Water goes in the bottom, out the top.*

- *Setups are three-dimensional: Flasks can rotate toward you, hoses upward, and so on.*

Separating or purifying liquids by vaporization and condensation is a very important step in one of our oldest professions. The word "still" lives on as a tribute to the importance of organic chemistry. There are two important points here:

1. *Vaporization.* Turning a liquid to a vapor.
2. *Condensation.* Turning a vapor to a liquid.

Remember these. They show up on quizzes.

But when do I use distillation? That is a very good question. Use the guidelines below to pick your special situation, and turn to that section. But you *should* read *all* of the sections anyway.

1. *Class 1: Simple distillation.* Separating liquids boiling below 150°C at one atmosphere (1 atm) from:
   a. Nonvolatile impurities.
   b. Another liquid that boils at least 25°C higher than the first. The liquids should dissolve in each other.
2. *Class 2: Vacuum distillation.* Separating liquids that boil above 150°C at 1 atm from:
   a. Nonvolatile impurities.
   b. Another liquid boiling at least 25°C higher than the first. They should dissolve in one another.
3. *Class 3: Fractional distillation.* Separating liquid mixtures, soluble in each other, that boil at less than 25°C from each other at 1 atm.
4. *Class 4: Steam distillation.* Isolating tars, oils, and other liquid compounds that are *insoluble,* or slightly soluble, *in water at all temperatures.* Usually, natural products are steam distilled. They do *not* have to be liquids at room temperatures. (For example, caffeine, a solid, can be isolated from green tea.)

Remember, these are guides. If your compound boils at 150.0001°C, don't whine that you must do a vacuum distillation or both you and your product will die. I expect you to have some judgment and to pay attention to your instructor's specific directions.

DISTILLATION NOTES

1. Except for Class 4, steam distillation, two liquids that are to be separated must dissolve in each other. If they did not, they would form separable layers, which you could separate in a separatory funnel (see Chapter 15, "Extraction and Washing").
2. Impurities can be either **soluble** or **insoluble.** For example, the material that gives cheap wine its unique bouquet is soluble in the alcohol. If you distill cheap wine, you get clear grain alcohol separated from the "impurities," which are left behind in the distilling flask.

CLASS 1: SIMPLE DISTILLATION (FIG. 19.1)

For separation of liquids that boil below 150°C at 1 atm from:

1. Nonvolatile impurities.
2. Another liquid that boils 25°C higher than the first liquid. *They must dissolve in each other.*

Sources of Heat

If one of the components boils below 70°C and you use a Bunsen burner, you may have a hard time putting out the fire. Use a steam bath or a heating mantle. Different distillations will require different handling (see Chapter 17, "Sources of Heat"). All distillations always require heating, so Chapter 17 is really closely tied to this section. This goes for enlightenment on using **boiling stones** and **clamps** as well (Chapter 18, "Clamps and Clamping").

FIGURE 19.1   A complete, entire simple distillation setup.

Product escapes and poisons everyone

Classic loose joint

FIGURE 19.2   The "commonly-camouflaged-until-it's-too-late" open joint.

The Three-Way Adapter

If there is any *one place* your setup will fall apart, here it is (Fig. 19.2). When you set up the jointware, it is important that all the joints *line up*. This is tricky, since, as you push one joint together, another pops right out. If you're not sure, call your instructor to inspect your work. Remember,

All joints must be tight!

The Distilling Flask

Choose a distilling flask carefully. If it's too big, you'll lose a lot of your product. If it's too small, you might have to distill in parts. *Don't fill the distilling flask more than half full.* Less than one-third full and you'll probably lose product. More than one-half full and you'll probably have undistilled material thrown up into the condenser (and into your previously clean product). Fill the distilling flask with the liquid you want to distill. You can remove the thermometer and thermometer adapter, fill the flask using a funnel, and then put the thermometer and its adapter back in place.

If you're doing a **fractional distillation** with a **column** (a Class 3 distillation), you should've filled the flask *before* clamping the setup. (Don't ever pour your mixture down a column. That'll contaminate everything!) You'll just have to disassemble some of the setup, fill the flask, reassemble what you've taken down, and pray that you haven't knocked all of the other joints out of line.

Put in a boiling stone if you haven't already. These porous little rocks promote bubbling and keep the liquid from superheating and flying out of the flask. This flying around is called **bumping.** Never drop a boiling stone into hot liquid, or you may be rewarded by having your body soaked in the hot liquid as it foams out at you.

Make sure all the joints in your setup are tight. Start the heat s-l-o-w-l-y until gentle boiling begins and liquid starts to drop into the receiving flask at the rate of about 10 drops per minute. *This is important.* If nothing comes over, you're not distilling, but merely wasting time. You may have to turn up the heat to keep material coming over.

The Thermometer Adapter

Read all about it. Ways of having fun with thermometer adapters have been detailed (see text accompanying Figs. 4.7–4.11 in Chapter 4, "Jointware").

The Ubiquitous Clamp

A word about clamps. *Use!* They can save you $68.25 in busted glassware (see Chapter 18, "Clamps and Clamping").

The Thermometer

Make sure the entire thermometer bulb is *below the side arm of the three-way adapter.* If you don't have liquid droplets condensing on the thermometer bulb, the temperature you read is *nonsense.* Keep a record of the temperature of the liquid or liquids that are distilling. It's a check on the purity. Liquid collected over a 2°C range is fairly pure. Note the similarity of this range with that of the *melting point* of a pure compound (see Chapter 12, "The Melting-Point Experiment").

The Condenser

Always keep cold water running through the condenser, enough so that *at least the lower half is cold* to the touch. Remember that water should go *in the bottom* and *out the top* (Fig. 19.1). Also, the water pressure in the lab may change from time to time and usually goes up at night, since little water is used then. So, if you are going to let condenser cooling water run overnight, tie the tubing on at the condenser and the water faucet with wire or something. And if you don't want to flood out the lab, see that the outlet hose can't flop out of the sink.

The Vacuum Adapter

It is important that the tubing connector remain *open to the air;* otherwise, the entire apparatus will, quite simply, explode.

Warning: Do not just stick the vacuum adapter on the end of the condenser and hope that it will not fall off and break.

This is foolish. I have no sympathy for anyone who will not use clamps to save their own breakage fee. They deserve to pay.

The Receiving Flask

The receiving flask should be large enough to collect what you want. You may need several, and they may have to be changed during the distillation. Standard practice is to have one flask ready for what you are going to throw away and others ready to save the stuff you want to save.

The Ice Bath

Why everyone insists on loading up a bucket with ice and trying to force a flask into this mess, I'll never know. How much cooling do you think you're going to get with just a few small areas of the flask barely touching ice? Get a suitable receptacle—a large beaker, enameled pan, or whatever. *It should not leak.* Put it under the flask. Put some water in it. *Now add ice.* Stir. Serves four.

Ice bath really means ice-water bath.

THE DISTILLATION EXAMPLE

Say you place 50 mL of liquid A (bp 50°C) and 50 mL of liquid B (bp 100°C) in a 250-mL round-bottom flask. You drop in a boiling stone, fit the flask in a distillation setup, and turn on the heat. Bubbling starts, and soon droplets form on the thermometer bulb. The temperature shoots up *from room temperature to about 35°C,* and a liquid condenses and drips into the receiver. That's bad. The temperature should be close to 50°C. This low-boiling material is the **forerun** of a distillation, and you won't want to keep it.

Keep letting liquid come over until the temperature stabilizes at about 49°C. Quick! Change receiving flasks now!

The new receiving flask is on the vacuum adapter, and the temperature is about 49°C. Good. Liquid comes over, and you heat to get a rate of about 10 drops per minute collected in the receiver. As you distill, the temperature slowly increases to maybe 51°C and then starts moving up rapidly.

Here you stop the distillation and change the receiver. Now in one receiver you have a *pure liquid, bp 49–51°C.* Note this **boiling range.** It is just as good a test of purity as a melting point is for solids (see Chapter 12, "The Melting-Point Experiment").

Always report a boiling point for liquids as routinely as you report melting points for solids. The boiling point is actually a **boiling range** and should be reported as such:

"bp 49–51°C"

Now, if you put on a new receiver and start heating again, you may discover *more material coming over at 50°C!* Find that strange? Not so. All it means is that you were distilling too rapidly and some of the low-boiling material was left behind. It is very difficult to avoid this situation. Sometimes it is best to ignore it, unless a yield is very important. You can combine this "new" 50°C fraction with the other good fraction.

For liquid B, boiling at 100°C, merely substitute some different boiling points and go over the same story.

THE DISTILLATION MISTAKE

OK, you set all this stuff up to do a distillation. Everything's going fine. Clamps in the right place. No arthritic joints, even the vacuum adapter is clamped on, and the thermometer is at the right height. There's a bright golden haze on the meadow, and everything's going your way. So, you begin to boil the liquid. You even remembered the boiling stone. Boiling starts slowly, then more rapidly. You think, "This is *it!*" Read that temperature, now. Into the notebook, "The mixture started boiling at 26°C."

And you are dead wrong.

What happened? Just ask—

Is there liquid condensing on the thermometer bulb??
No!

So, congratulations, you've just recorded the room temperature. There are days when more than half of the class will report distillation temperatures as "Hey, I see it starting to boil now" temperatures. Don't participate. Just keep watching as the liquid boils. Soon, droplets *will* condense on the thermometer bulb. The temperature will go up quickly, and then *stabilize.* Now read the temperature. That's the boiling point. But wait! It's *not* a distillation temperature until that first drop of liquid falls into the receiving flask.

Just before you leave for your next class, **vacuum distillation,** you might want to check out Chapter 34, "Theory of Distillation."

CLASS 2: VACUUM DISTILLATION

For separation of liquids that boil above 150°C at 1 atm from:

1. Nonvolatile impurities.

Screw clamp

Air or gas inlet

Clear area shows
grease on all joints

Inlet tube

To vacuum source

Fine capillary

FIGURE 19.3 A vacuum distillation setup.

2. Another liquid that boils 25°C higher than the first liquid. They must dissolve in each other. This is like the **simple distillation** with the changes shown (Fig. 19.3).

Why vacuum distill? If the substances boil at high temperatures at 1 atm, they may decompose when heated. Putting the liquid under vacuum makes the liquid boil at a lower temperature. With the pressure reduced, there are fewer molecules in the way of the liquid you are distilling. Since the molecules require less energy to leave the surface of the liquid, *you can distill at a lower temperature,* and *your compound doesn't decompose.*

Pressure Measurement

If you want to measure the pressure in your vacuum distillation setup, you'll need a **closed-end manometer.** There are a few different types, but they all work essentially in the same way. I've chosen a "stick" type (Fig. 19.4). This particular model needs help from a short length of rubber tubing and a glass T to connect it to the vacuum distillation setup.

1. Turn on the source of vacuum and wait a bit for the system to stabilize.

2. Turn the knob on the manometer so that the notch in the joint lines up with the inlet.

3. Wait for the mercury in the manometer to stop falling.

FIGURE 19.4 A closed-end "stick" manometer.

4. Read the *difference* between the inner and outer levels of mercury. This is the system pressure, literally in **millimeters of mercury,** which we now call *torr.*

5. Turn the knob on the manometer to disconnect it from the inlet. Don't leave the manometer permanently connected. Vapors from your distillation, water vapor from the aspirator, and so on, may contaminate the mercury.

Manometer Hints

1. Mercury is toxic, the vapor from mercury is toxic, spilled mercury breaks into tiny globules that evaporate easily and are toxic, it'll alloy with your jewelry, and so on. Be very careful not to expose yourself (or anyone else) to mercury.

2. If the mercury level in the *inner* tube goes *lower than that of the outer tube*, it does not mean that you have a negative vacuum. Some air or other vapor has gotten into the inner stick, and with the vacuum applied, the vapor expands and drives the mercury in the inner tube lower than that in the outer tube. This manometer is unreliable, and you should seek a replacement.

3. If a rubber tube connected to the vacuum source and the system (or manometer) *collapses,* you've had it. The system is no longer connected to the vacuum source, and as air from the bleed tube or vapor from the liquid you're distilling fills your distillation setup, the pressure in the system goes up. Occasionally test the vacuum hoses, and if they collapse under vacuum, replace them with sturdier hoses that can take it. You can test the tubing by connecting one end to the water aspirator, pinching the other end closed, and turning the water on so that it comes through the water aspirator full blast. If the tubing collapses now, it'll collapse during the vacuum distillation.

Leaks

Suppose, by luck of the draw, you've had to prepare and purify 1-octanol (bp 195°C). You know that if you simply distill 1-octanol, you run the risk of decomposing it, so you set up a vacuum distillation. You hook your setup to a water aspirator and water trap, and you attach a closed-end "stick" manometer. You turn the water for the aspirator on full blast and open the stick manometer. After a few minutes, nothing seems to be happening. You *pinch* the tubing going to the vacuum distillation setup (but not to the manometer), closing the setup off from the source of vacuum. Suddenly, the mercury in the manometer starts to drop. You *release* the tube going to the vacuum distillation setup, and the mercury jumps to the upper limit. You have **air leaks** in your vacuum distillation setup.

Air leaks can be difficult to find. At best, you push some of the joints together again and the system seals itself. At worst, you have to take apart all the joints and regrease every one. Sometimes you've forgotten to grease all the joints. Often a joint has been etched to the point that it cannot seal under vacuum, though it is perfectly fine for other applications. Please get help from your instructor.

Pressure and Temperature Corrections

You've found all the leaks, and the pressure in your vacuum distillation setup is, say, 25 torr. Now you need to know the boiling point of your compound,

Observed bp

FIGURE 19.5 One-point conversion.

1-octanol, this time at 25 torr, not 760 torr. You realize it'll boil at a lower temperature, but just how low? The handy nomographs can help you estimate the new boiling point.

This time you have the boiling point at 760 torr (195°C) and the pressure you are working at (25 torr). Using Fig. 19.5, you:

1. Find the boiling point at 760 torr (195°C) on line B (the middle one).
2. Find the pressure you'll be working at (25 torr) on line C (the one on the far right). You'll have to estimate this point.
3. Using a straightedge, line up these two points and see where the straightedge cuts the observed boiling point line (line A, far left). I get about 95°C.

So, a liquid that boils at 195°C at 760 torr will boil at about 95°C at 25 torr. Remember, this is an estimate.

Now suppose you looked up the boiling point of 1-octanol and all you found was: 98[19]. This means that the boiling point of 1-octanol is 98°C at 19 torr. Two things should strike you.

1. This is a *higher* boiling point at a *lower* pressure than we've gotten from the nomograph.

2. I wasn't kidding about this process being an estimation of the boiling point.

Now we have a case of having an observed boiling point at a pressure that is *not* 760 torr (1-octanol again; 98°C at 19 torr). We'd like to get to 25 torr, our working pressure. This requires a double conversion, as shown in Fig. 19.6.

1. On the observed bp line (line A), find 98°C.

2. On the pressure-in-torr line (line C), find 19.

Observed bp

FIGURE 19.6   Two-point conversion.

3. Using a straightedge, connect those points. Read the bp corrected to 760 torr (line B): I get 210°C.

4. Now, using the 210°C point as a fulcrum, pivot the straightedge until the 210°C point on line B and the pressure you're working at (25 torr) on line C line up. You see, you're in the same position as in the previous example with a "corrected to 760 torr bp" and a working pressure.

5. See where the straightedge cuts the observed boiling point line (line A). I get 105°C.

So, we've estimated the boiling point at about 105°C at 25 torr. The last time it was 95°C at 25 torr. Which is it? Better you should say you *expect* your compound to come over at 95–105°C. Again, this is not an unreasonable expectation for a vacuum distillation.

The pressure–temperature nomograph (Fig. 19.7) is really just a simple, graphical application of the Clausius–Clapeyron equation. If you know the heat of

Observed bp

FIGURE 19.7 A clean nomograph for your own use.

vaporization of a substance, as well as its normal boiling point, you can calculate the boiling point at another temperature. You do have to assume that the heat of vaporization is constant over the temperature range you're working with, and that's not always so. Where's the heat of vaporization in the nomograph? One is built into the slopes and spacings on the paper. And, yes, that means that the heat of vaporization is forced to be the same for all compounds, be they alkanes, aldehydes, or ethers. So do not be surprised at the inaccuracies in this nomograph; be amazed that it works as well as it does.

Vacuum Distillation Notes

1. Read all of the notes on Class 1 (simple distillation).

2. The thermometer can be replaced by a **gas inlet tube.** It has a long, fine capillary at one end (Fig. 19.3). This helps to stop the *extremely bad bumping* that goes along with vacuum distillations. The fine stream of bubbles through the liquid produces the same results as a boiling stone. Boiling stones are useless, since all the adsorbed air is whisked away by the vacuum and the nucleating cavities plug up with liquid. The fine capillary does not let in a lot of air, so we are doing a vacuum distillation anyway. Would you be happier if I called it a **reduced-pressure** distillation? An inert gas (nitrogen?) may be let in if the compounds decompose in air.

3. If you can get a **magnetic stirrer** and **magnetic stirring bar,** you won't have to use the gas inlet tube approach. Put a magnetic stirring bar in the flask with the material you want to vacuum distill. Use a *heating mantle* to heat the flask, and put the magnetic stirrer under the mantle. When you turn the stirrer on, a magnet in the stirrer spins, and the stirring bar (a Teflon-coated magnet) spins. Admittedly, stirring through a heating mantle is not easy, but it can be done. *Stirring the liquid also stops the bumping.*

> **Remember, first the stirring, then the vacuum, then the heat—or WHOOSH! Got it?**

4. Control of heating is *extremely critical.* I don't know how to shout this loudly enough on paper. Always apply the vacuum first and watch the setup for awhile. Air dissolved or trapped in your sample or a highly volatile leftover (maybe ethyl ether from a previous extraction) can come flying out of the flask *without the heat.* If you heated such a setup a bit and then applied the vacuum, your sample would blow all over, possibly right into the receiving flask. Wait for the contents of the distilling flask to calm down before you start the distillation.

5. If you know you have low-boiling material in your compound, think about distilling it at atmospheric pressure first. If, say, half the liquid you want to vacuum distill is ethyl ether from an extraction, consider doing a simple distillation to get rid of the ether. Then the ether (or any other low-boiling compound) won't be around to cause trouble during the vacuum distillation. If you distill first at

1 atm, *let the flask cool before you apply the vacuum*. Otherwise, your compound will fly all over and probably will wind up, undistilled and impure, in your receiving flask.

6. *Grease all joints, no matter what* (see "Greasing the Joints" in Chapter 4). Under vacuum, it is easy for any material to work its way into the joints and turn into concrete, and the joints will never, ever come apart again.

7. The **vacuum adapter** is connected to a **vacuum source,** either a **vacuum pump** or a **water aspirator.** Real live vacuum pumps are expensive and rare and not usually found in the undergraduate organic laboratory. If you can get to use one, that's excellent. See your instructor for the details. The **water aspirator** is used lots, so read up on it (see Chapter 13).

8. During a vacuum distillation, it is not unusual to collect a *pure compound over a 10–20°C temperature range.* If you don't believe it, you haven't ever done a vacuum distillation. It has to do with pressure changes throughout the distillation because the setup is far from perfect. Although a vacuum distillation is not difficult, it requires peace of mind, large quantities of patience, and a soundproof room to scream in so you won't disturb others.

9. A **Claisen adapter** in the distilling flask lets you take temperature readings and can help stop your compound from splashing over into the distillation receiver (Fig. 19.8). On a microscale, though, the extra length of the Claisen adapter could hold up a lot of your product. Check with your instructor. You could also use a **three-neck flask** (Fig. 19.9). Think! And, of course, use some glassware too.

CLASS 3: FRACTIONAL DISTILLATION

For separation of liquids that are soluble in each other and boil fewer than 25°C from each other, use fractional distillation. This is like simple distillation with the changes shown (Figs. 19.10 and 19.11).

Fractional distillation is used when the components to be separated boil *within 25°C of each other.* Each component is called a **fraction.** Clever where they get the name, eh? This temperature difference is not gospel. And don't expect terrific separations, either. Let's just leave it at *close boiling points.* How close? That's hard to answer. "Is an orange?" That's easier to answer. If the experiment tells you to "fractionally distill," at least you'll be able to set it up right.

How This Works

If one distillation is good, two are better. And fifty, better still. So you have lots and lots of little, tiny distillations occurring on the surfaces of the **column packing,** which can be glass beads, glass helices, ceramic pieces, metal chips, or even stainless steel wool.

Claisen adapter

FIGURE 19.8   A Claisen adapter so you can vacuum distill and take temperatures, too.

As you heat your mixture, it boils, and the vapor that comes off this liquid is *richer in the lower-boiling component.* The vapor moves out of the flask and condenses, say, on the first centimeter of column packing. Now, the composition of the liquid still in the flask has changed a bit—it is *richer in the higher-boiling component.* As more of *this* liquid boils, more hot vapor comes up, mixes with the first fraction, and produces a new vapor of different composition—*richer yet in the more volatile (lower-boiling) component.* And guess what? This new vapor condenses in the *second centimeter* of column packing. And again, and again, and again.

Now all of these are **equilibrium steps.** It takes some time for the fractions to move up the column, get comfortable with their surroundings, meet the neighbors . . . . And if you *never* let any of the liquid–vapor mixture out of the column, a condition called **total reflux,** you might get a single pure component at the top—namely, the lower-boiling, more volatile component all by itself! This is an ideal separation.

One neck stoppered

FIGURE 19.9 Multipurpose setup with a three-neck flask.

Fat lot of good that does you when you have to hand in a sample. So, you turn up the heat, let some of the vapor condense, and *take off this top fraction.* This raises hell in the column. *Nonequilibrium conditions abound—mixing. Arrrgh!* No more completely pure compound. And the faster you distill, the faster you let material come over, the higher your **throughput**—the worse this gets. Soon you're at **total takeoff,** and there is no time for an equilibrium to become established. And if you're doing that, you shouldn't even bother using a column.

You must strike a compromise. Fractionally distill *as slowly as you can,* keeping in mind that eventually the lab does end. Slow down your fractional distillations; I've found that 5–10 drops per minute coming over into the receiving flask is usually suggested. It will take a bit of practice before you can judge the best rate for the best separation. See your instructor for advice.

Fractional Distillation Notes

1. Read all of the notes on Class 1 (simple distillation).

2. Make *sure* you have not confused the **column** with the **condenser**. The *column is wider and has glass projections inside* at the bottom to hold up the packing.

3. *Don't break off the projections!*

FIGURE 19.10 The fractional distillation setup.

4. Do not run water through the jacket of the column!

5. Sometimes, the column is used *without* the column packing. This is all right, too.

6. If it is necessary, and it usually is, push a wad of heavy metal wool down the column, *close to the support projections,* to support the packing chips.

Unjacketed column

Stopper

Column packing

Glass projections hold up column packing

Claisen tube here increases efficiency

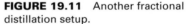
Heat source

FIGURE 19.11 Another fractional distillation setup.

Sometimes the packing is entirely this stainless steel wool. You can see that it is self-supporting.

7. Add the column packing. Shake the column lightly to make sure none of the packing will fall out later into your distillation flask.

8. With all the surface area of the packing, a lot of liquid is *held up* on it. This phenomenon is called **column holdup,** *since it refers to the material retained in the column. Make sure you have enough compound to start with, or it will all be lost on the packing.*

9. A **chaser solvent** or **pusher solvent** is sometimes used to help blast your compound off the surface of the packing material. It should have a *tremen-*

dously high boiling point relative to what you were fractionating. After you've collected most of one fraction, some of the material is left on the column. So, you throw this chaser solvent into the distillation flask, fire it up, and start to distill the chaser solvent. As the chaser solvent comes up the column, it heats the packing material, your compound is blasted off the column packing, and more of your compound comes over. Stop collecting when the temperature starts to rise—that's the chaser solvent coming over now. As an example, you might expect *p*-xylene (bp 138.4°C) to be a really good chaser, or pusher, for compounds that boil at less than, say, 100°C.

But you have to watch out for the deadly azeotropes.

AZEOTROPES

Once in a while, you throw together two liquids and find that you cannot separate part of them. And I don't mean because of poor equipment, or poor technique, or other poor excuses. You may have an **azeotrope,** a mixture with a *constant boiling point.*

One of the best-known examples is the ethyl alcohol–water azeotrope. This 96% alcohol–4% water solution will boil to dryness, at a *constant temperature.* It's slightly scary, since you learn that a liquid is a pure compound if it boils at a constant temperature. And you thought you had it made.

There are two types of azeotropes. If the azeotrope boils off first, it's a **minimum-boiling azeotrope.** After it's all gone, if there is any other component left, only then will that component distill.

If any of the components come off first, and then the azeotrope, you have a **maximum-boiling azeotrope.**

Quiz question:

Fifty milliliters of a liquid boils at 74.8°C from the beginning of the distillation to the end. Since there is no wide boiling range, can we assume that the liquid is pure?

No. It may be a constant-boiling mixture called an azeotrope.

You should be able to see that you have to be really careful in selecting those chaser or pusher solvents mentioned. Sure, water (bp 100°C) is hot enough to chase ethyl alcohol (bp 78.3°C) from any column packing. Unfortunately, water and ethyl alcohol form an azeotrope and the technique won't work. (*Please* see Chapter 34, "Theory of Distillation.")

CLASS 4: STEAM DISTILLATION

Mixtures of tars and oils must not dissolve well in water (well, not much, anyway), so we can steam distill them. The process is pretty close to simple distillation, but you should have a way of getting *fresh hot water into the setup* without stopping the distillation.

Why steam distill? If the stuff you're going to distill is *only slightly soluble in water* and may decompose at its boiling point and the bumping will be terrible with a vacuum distillation, it is better to **steam distill.** Heating the compound in the presence of steam makes the compound boil at a lower temperature. This has to do with partial pressures of water and organic oils and such.

There are two ways of generating steam: externally and internally.

External Steam Distillation

In an **external steam distillation,** you lead steam from a steam line through a water trap, and thus into the system. The steam usually comes from a **steam tap** on the bench top. This is classic. This is complicated. This is dangerous.

1. Set up your external steam distillation apparatus in its entirety. Have *everything* ready to go. This includes having the material you want to distill in the distilling flask, the steam trap already attached, condensers up and ready, a large receiving flask, and so on. All you should have to do is attach a single hose from the steam tap to your steam trap and start the steam.

2. *Have your instructor check your setup before you start!* I cannot shout this loudly enough on this sheet of paper. Interrupting an external steam distillation, just because you forgot your head this morning, is a real trial.

3. Connect a length of rubber tubing to your bench steam outlet, and lead the rubber tubing into a drain.

4. Now, watch out! Slowly, carefully, open the steam stopcock. Often you'll hear clanging, bonking, and thumping, and a mixture of rust, oil, and dirt-laden water will come spitting out. Then some steam bursts come out, and, finally, you have a stream of steam. Congratulations. You have just *bled the steam line.* Now close the steam stopcock, wait for the rubber tubing to cool a bit, and then . . .

5. Carefully (**Caution**—may be hot!) attach the rubber tubing from the steam stopcock to the inlet of your steam trap.

6. Open the steam trap drain, then carefully reopen the bench steam stopcock. Let any water drain out of the trap, then carefully close the drain clamp. Be careful.

7. You now have steam going through your distillation setup, and as soon as product starts to come over, you'll be doing an external steam distillation. Periodically, open the steam trap drain (**Caution**—hot!), and let the condensed steam out.

8. Apparently you can distill as fast as you can let the steam into your setup, *as long as all of the steam condenses* and doesn't go out into the room. Sometimes when you steam distill, you need to hook two condensers together, making a very long supercondenser. Check with your instructor.

9. When you're finished (see "Steam Distillation Notes" following), turn off the steam, let the apparatus cool, and dismantle everything.

Steam in

Screw clamp

Steam distillation in three-neck flask

To drain

FIGURE 19.12   One example of an external steam trap.

You can use many types of steam traps with your distillation setup. I've shown two (Figs. 19.12 and 19.13), but these are not the only ones, and you may use something different. The point is to note the **steam inlet** and the **trap drain** and how to use them.

Internal Steam Distillation

1. You can add hot water to the flask (Fig. 19.14) that will generate steam and thus provide an **internal source of steam.** This method is used almost exclusively in an undergraduate organic lab for the simple reason that it is so simple.

2. Add to the distilling flask at least three times as much water (maybe more) as sample. Do not fill the flask much more than half full (three-quarters, maybe). You've got to be careful. Very careful.

3. Periodically, add more *hot water* as needed. When the water boils and turns to steam, it also leaves the flask, carrying product.

STEAM DISTILLATION NOTES

1. Read all of the notes on Class 1 distillations.

2. Collect some of the distillate, the stuff that comes over, in a small test tube. Examine the sample. If you see *two layers,* or *if the solution is cloudy,* you're not done. Your product is still coming over. Keep distilling and keep adding hot water to generate more steam. *If you don't see any layers, don't assume you're done.* If the sample is slightly soluble in the water, the two layers or cloudiness might not show up. Try *salting out.* This has been mentioned before in

FIGURE 19.13 Another example of an external steam trap.

connection with **extraction** and **recrystallization** as well (see "Salting Out" in Chapter 13 and "Extraction Hints" in Chapter 15). Add some salt to the solution you've collected in the test tube, shake the tube to dissolve the salt, and, if you're lucky, more of your product may be squeezed out of the aqueous layer, forming a *separate layer*. If that happens, *keep steam distilling* until the product does not come out when you treat a test solution with salt.

3. There should be two layers of liquid in the receiving flask at the end of the distillation. One is *mostly water;* the other is *mostly product.* Get a small (10 × 75-mm) test tube, and add about 0.5–1.0 mL water. Now add 2–4 drops of the layer you *think* is aqueous, and swirl the tube. If the stuff *doesn't dissolve* in

FIGURE 19.14 An internal steam distillation setup.

the water, it's *not* an aqueous (water) layer. The stuff may sink to the bottom, float on the top, do *both,* or even *turn the water cloudy!* It will *not,* however, dissolve.

4. If you have to get more of your organic layer out of the water, you can do a **back-extraction** with an immiscible solvent (see "The Road to Recovery— Back-Extraction" in Chapter 15).

SIMULATED BULB-TO-BULB DISTILLATION: FAKELROHR

If you have to distill small amounts and you don't want to lose a lot of product on the surface of condensers, adapters, and all that, you might use a Kugelrohr apparatus (Fig. 19.15). The crude material, in a flask in an oven, distills from bulb-to-bulb

FIGURE 19.15 Simple schematic of a Kugelrohr bulb-to-bulb distillation apparatus.

FIGURE 19.16 Fakelrohr setups: (*a*) jointware, (*b*) microscale, and (*c*) Williamson test tube.

with very little loss. As a bonus, the entire stack of bulbs (and flask) rotate à la rotary evaporator (see "The Rotary Evaporator," Chapter 21), so the material you're distilling coats the flask and distills off the much larger surface of the flask walls. Nice, but expensive, and with a lab full of people, quite a holdup if all you have is one unit.

What to do? Well, we just have to fake it. You connect two flasks by a vacuum adapter and distilling head, and voilà: Fakelrohr. OK, it's not true bulb-to-bulb distillation, and they don't rotate, but it's the shortest-path distillation we can set up in the undergraduate organic laboratory outside of the microscale Hickman still (see "Microscale Distillation," Chapter 20). Fakelrohr setups for jointware, microscale jointware, and Williamson test tubeware are shown in Figure 19.16.

EXERCISES

1. Often a clear product, upon distillation, initially gives a cloudy liquid. This observation has been called a "reverse steam distillation." Explain.

2. Define boiling point.

3. In a vacuum distillation, you have to start the stirring, apply the vacuum, and then apply the heating. What would happen in a vacuum distillation if you followed these instructions in an another order?

4. Why is a large reflux ratio (see "Theory of Distillation," Chapter 35) considered desirable for efficient fractionation with a column?

5. Should you add boiling stones to reduce bumping in a vacuum distillation?

6. Why use a water trap in a vacuum distillation? What makes more sense, connecting the side outlet of the flask to the vacuum source and the center tube to the system, or vice versa?

7. Explain a little bit about how both a maximum-boiling and a minimum-boiling azeotropic mixture would behave throughout distillations from both sides of the azeotropic temperature.

8. If the heat source in a distillation is a heating mantle, why should it be set in a holder so it isn't resting on the bench top? Would a stirring hot plate have to be held above the bench top? Why must the distilling flask be clamped so it cannot fall? Imagine some problem occurring that requires you to immediately remove the heat from the distilling flask. Think about this, then go back to the other two questions.

9. You don't normally use a plastic Keck clip to hold the distilling flask on the rest of the setup. Why not? How hard is it to chip a melted Keck clip off a jointware flask, anyway?

MICROSCALE DISTILLATION

LIKE THE BIG GUY

Before you read this chapter, read up on "Class 1: Simple Distillation" (Chapter 19), so you'll know what I'm talking about, so you can set up the same apparatus and do the same distillation for the same classes you've always done.

That's not completely true. Using this apparatus at the true microscale (0.01 mole), you have two drops of a liquid to distill. I'd see my instructor before I did any of these techniques with fewer than 10 mL.

Class 1: Simple Distillation

Perhaps you'll be using a conical vial rather than a round-bottom flask, but the setup and operation are the same as with the larger apparatus. *Note:* The tubing carrying cooling water can be so unwieldy that moving the tubing can pull this tiny setup out of alignment.

Class 2: Vacuum Distillation

Because the setup is so small and your sample size is so small, you've got to be very careful with vacuum distillation. With ᵀ 14/20 ware (it's a little bigger), the operations are identical to those with the ᵀ 24/40 or ᵀ 19/22 sizes. I don't *have* the ᵀ 14/10 setup yet, so I can only extrapolate. You'll have the same entertaining experiences. Watch the vacuum hose, too. It can be so unwieldy that it pulls the apparatus out of line if it's moved.

Oh: Check out "The O-Ring Cap Seal" discussion for a perspective on microscale vacuum distillation (see Chapter 5, "Microscale Jointware").

Class 3: Fractional Distillation

You'll probably substitute one of those **air condensers** for the distilling column. As a straight, unjacketed tube, the air condenser is not very efficient. I'd be careful about packing it with anything; the column holdup, as small as it is, is likely to hold up a lot of your product.

Class 4: Steam Distillation

With sample sizes this small, internal steam distillation is the only way out. With the microscale equipment, you can inject water by syringe rather than opening up the apparatus to pour water in, and lose less product. Again, if your sample is very small, see your instructor about any problems.

MICROSCALE DISTILLATION II: THE HICKMAN STILL

If you have only 1–2 mL to distill, then the Hickman still is probably your only way out (Fig. 20.1).

The Hickman Still Setup

1. Set up the stirring hot plate and sand bath. I suppose you *could* stick a thermometer in the sand to give you an idea of how warm the sand is.

2. Put the sample you want to distill into an appropriately sized conical vial. Two things: Don't fill the vial more than halfway, and don't use such a big vial that the product disappears. Less than one-third full, and you'll lose a lot on the walls of the vial.

FIGURE 20.1 Two different Hickman still setups.

3. These vials can fall over easily. Keep them in a holder, or clamp them until you're ready to use them.

4. Add a microboiling stone *or* a spin vane, not both, to the vial.

5. Get a clamp ready to accept the Hickman still at about the position you want, just open and loosely clamped—horizontally—to the ring stand.

6. Scrunch the conical vial down into the sand, and push it to where you want it to be. Hold the vial at the top.

7. Put the Hickman still on the vial. Now, even though you have a sand buttress, this baby is still top-heavy and can fall. Hold the vial *and* the still somewhere about where they meet.

8. Jockey the clamp and/or the vial and still so that they meet. Clamp the still at the top in the clamp jaws. Get the vial and still in their final resting place; tighten the clamp to the ring stand.

9. Hang a thermometer by a clamp and stopper so that the thermometer goes straight down the throat of the still. Place the thermometer bulb at about the joint of the vial and still. Do *not* let the thermometer touch the sides of the vial or still.

10. You're ready to heat.

Hickman Still Heating

You don't have very much liquid here, so it doesn't take much heat to boil your sample. If you have no idea what temperature your sample will boil at, you should just heat carefully until you see it boil, and then *immediately* turn the heat down 10–20%. Of course, you have a microboiling stone or magnetic spin vane in there, don't you?

As you heat the sample, you'll see a ring of condensate travel up the vial, and when it hits the thermometer bulb, the temperature reading should move up. As usual, if there's **refluxing liquid** on the bulb and the temperature has stabilized, that's the boiling point.

Then, the vapor travels up into the still head, condenses on the glass surface, and falls down to be caught in the bulge of the still (Fig. 20.1). When you feel you've collected enough liquid (this will depend on the experiment), you can go after it with a pipet and rubber bulb.

Recovering Your Product

If you have a thermometer down the throat of the **Hickman still,** forget about getting your product out with a disposable pipet; there's just not enough room, period. If you use a $5\frac{3}{4}$-in. pipet, you'll likely snap the tip off in the still when you try to bend it to reach your product that is caught in the bulge of the still. Yes, you *can* just about wangle the end of a 9-in. pipet into the crevice and pick up your product, but there's a good chance of breaking the tip off anyway. You might wind up having to put a bend in the tip by using a burner and your considerable glassworking skills. You have to use a small burner and a small flame and heat the tube carefully so it doesn't close off or melt off—in short, see your instructor about this. You should wind up with a slightly bent tip at the end of the 9-in. pipet. What I haven't yet tried, and wonder why it hasn't been done yet, is to put a small length of the gas-collecting tubing—very thin, narrow plastic or Teflon tubing—on the end of a pipet and bend it into a curve so you can go after your distillation product without resorting to glassworking. Again, the 9-in. pipet *can* just make it, but be very careful.

THE ROTARY EVAPORATOR

If your product is dissolved in a solvent, you can distill off the solvent, boil off the solvent, or gently evaporate off the solvent using a **rotary evaporator.** Everybody calls it a **rotovap** and uses the word as a verb, too: "*Rotovap* the solution to dryness." Generally, they mean a rotary evaporator made by the Büchi Corporation, so you might be told to "put it on the Büchi."

The key to the rotovap is that it spins your flask, held at an angle, under a vacuum, and the solvent continually coats the upper part of the flask with a thin film. This thin film of solvent, covering such a large area of the flask, then evaporates pretty quickly.

From the top, the Büchi rotovap (Fig. 21.1) consists of:

1. *Solvent-sucking inlet.* If you have a lot of solution, you can suck more of it way down in the rotating flask without missing a beat. With a **splash trap** on the rotovap, this feature is useless, however. Generally, you leave this unconnected and use it to let air into the rotovap by turning the stopcock at the top.

2. *Stopcock at the top.* Note how the handle points to the flow of the liquid or gas. Handle away from the inlet tube, stopcock closed; handle on the same side as the inlet tube, stopcock open.

FIGURE 21.1 The Büchi rotary evaporator.

196

3. *Condenser.* This is a pretty fancy condenser in which the cooling water flows through this long, coiled glass tube inside the condenser. Cool! This alone can account for the virtual disappearance of other styles of rotovap that are just as effective and considerably cheaper. They are just not as cool. There's a glass inlet and outlet tube (not shown in Figure 21.1) "around the back." They're close enough to the vacuum source connection that you might at first confuse the three, but a close inspection will reveal which is which.

4. *Vacuum connection.* This is where you, well, connect the vacuum source.

5. *Socket joint clip with screw lock.* This looks and acts a lot like a spring-type clothespin. Just like that kind of clothespin, you press the ends together to open the jaws of the clip and fit the jaws around the socket joint. Then, you turn a little knurled wheel to lock the jaws around the joint. To remove the clip, you may first have to unlock the jaws by unscrewing the little knurled wheel. If you don't know about this, when you come to remove the receiving flask by pressing the ends, the ends won't move. The first response here is not that the clip is broken, but that you have to unscrew the little knurled wheel so the jaws can open.

6. *§35/25 receiving flask.* Unlike T joints that can only rotate about their axis, ball-and-socket joints—or just socket joints—can rotate in almost any direction. An §35/25 joint has a diameter of 35 mm, with a bore of 25 mm going through it.

7. *On–off speed control.* What more needs to be said? Don't immediately crank this control all the way up. You can get to speeds that seem scary.

8. *Rotovap splash trap* (Fig. 21.2). If your solution bumps or foams, you can contaminate the condenser with whatever was dissolved in the solvent. Hence the trap.

9. *Your flask.* Besides not having it more than half-full, there's nothing to say here but this: Yes, you can use clips to hold the flask, and even the splash trap, on the rotovap, but I've always let the vacuum in the setup hold the flask and trap on. Just lucky, I guess.

10. *Heating bath.* Just that. Unless the bath is thermostatted, it's probably better to just put warm water in the bath and let it cool as the evaporation continues. As the liquid evaporates from the walls of the flask, as they say in the chemistry books,

To rotovap

Holes in glass tube let splashed or foaming liquid out and into trap—and not into the rotovap

To flask with solvent to be removed

FIGURE 21.2 A rotovap splash guard.

"the fraction of molecules with the highest kinetic energy" leave, and the solution you're rotovapping cools, which means you have to add more heat to keep things evaporating at the same rate. If you care to, after you've started your rotovap but before you raise the heating bath, feel the rotating flask. It'll get cold.

11. *Lab jack.* Raises and lowers the heating bath.

Rotovapping is like vacuum distillation, and in starting a vacuum distillation it's always first the stirring, then the vacuum, then the heat. Here stirring is replaced by spinning the flask. To start rotovapping a solution:

1. Make sure the sample-sucking inlet stopcock is closed off. It usually is; just make sure. Empty the receiving flask. Suppose the last person rotovapped off diethyl ether (bp 34.6°C) and didn't clean it out. Then you come by and want to rotovap off ethanol (bp 78.5°C). If ether is still in the receiving flask, as you pull a vacuum in the rotovap, the ether will start to reflux, filling the rotovap with ether vapor. Your ethanol molecules haven't a chance. They'll just sit there. So, clean out the receiving flask. Reattach the socket joint flask to the ball joint on the condenser, and use the special clip to hold the flask to the condenser outlet. If you're doing solvent recovery so you can reuse it later, it's best to have a single rotovap dedicated to a single solvent. Otherwise, you can get cross-contamination of one solvent in another. Cleaning and drying the receiving flask may not be enough. There's a lot of surface area in that condenser.

2. Start the cooling water flowing through the condenser. Gentle flow, please.

3. Make sure your round-bottom jointware flask fits the rotovap or you have adapters to make it fit.

4. Slowly, without hooking anything up yet, angle the flask so it is parallel to the joint on the rotovap. Did some of your solution spill out? A clear indication that your flask is too full. Half full is about the limit; a little less is better. Either switch to a bigger flask or you'll have to rotovap your solution in parts.

5. Without ever letting go of the flask (unless you've clipped it on), connect the flask to the joint on the rotovap. (Your flask might better have a splash trap connected to it, and *both* go onto the joint on the rotovap.) Still holding the flask, start the motor that spins the flask. Dial up a speed that you and your instructor are comfortable with.

6. Slowly establish a vacuum in the rotovap. When you are sure there's enough vacuum in the rotovap so the flask (and splash trap) won't drop off the beast, then, and only then, let go of the flask. Again, if you've clipped it on, you don't have to wait for the vacuum to build to hold the flask (and splash trap) on.

7. Jack up whatever heating bath is recommended, if any.

Now you let the flask spin and the solvent evaporate. It's a good idea to watch the setup every so often to see how far down the solvent is.

Again, rotovapping is like vacuum distillation, and in ending a vacuum distillation it's always first turn off the heat, then break the vacuum, then stop the stirring, and, again, stirring is replaced by spinning the flask. When you're ready to take the flask off the rotovap:

1. Lower or otherwise remove any heating bath.

2. Supporting the bottom of the flask (unless you've clipped it on), slowly let air into the rotovap; the sample-sucking inlet stopcock is the best place to do this.

3. Turn off the spinner motor.

4. Remove your flask.

5. Turn off whatever you used to establish a vacuum in the rotovap, either a water aspirator or a mechanical pump.

6. Turn off the cooling water.

7. Be a mensch, and empty the receiving flask for the next person. There might be a special container for these solvents. Check with your instructor.

8. The term *mensch* applies to persons of both genders. Look it up.

EXERCISES

1. Did you look up the Yiddish word *mensch*?

2. The last person to use the rotovap took off acetone. You need to take off ethanol. Should you check the receiving flask, and why should you? If a small amount of acetone is left in the receiving flask, is it a big deal or not? Why or why not?

REFLUX AND ADDITION

Just about 80% of the reactions in organic lab involve a step called **refluxing.** You use a reaction solvent to keep materials dissolved and at a constant temperature by boiling the solvent, condensing it, and returning it to the flask.

For example, say you have to heat a reaction to around 80°C for 17 hours. Well, you can stand there on your flat feet and watch the reaction all day. Me? I'm off to the **reflux.**

STANDARD REFLUX

Usually you'll be told what solvent to use, so selecting one should not be a problem. What happens more often is that you choose the reagents for your particular synthesis, put them into a solvent, and **reflux** the mixture. You boil the solvent and condense the solvent vapor *so that ALL of the solvent runs back into the reaction flask* (see "Class 3: Fractional Distillation" in Chapter 19). The *reflux temperature is near the boiling point of the solvent.* To execute a reflux,

1. Place the reagents in a round-bottom flask. The flask should be large enough to hold both the reagents and enough solvent to dissolve them, without being much more than half full.

2. You should now choose a solvent that
   a. Dissolves the reactants at the boiling temperature.
   b. Does *not* react with the reagents.
   c. Boils at a temperature that is high enough to cause the desired reaction to go at a rapid pace.

3. Dissolve the reactants in the solvent. Sometimes *the solvent itself is a reactant.* Then don't worry.

4. To stop bumping, add a boiling stone to the flask. A magnetic stirring bar and stirrer will work, too.

5. Place a condenser, upright, on the flask, connect the condenser to the water faucet, and run water through the condenser (Fig. 22.1). Remember: in at the bottom and out at the top.

6. Put a suitable heat source under the flask, and adjust the heat so that the solvent condenses *no higher than halfway up the condenser.* You'll have to stick around and watch for a while, since this may take some time to get started. Once the reaction is stable, though, go do something else. You'll be ahead of the game for the rest of the lab.

7. Once this is going well, leave it alone until the reaction time is up. If it's an overnight reflux, wire the water hoses on so they don't blow off when you're not there.

8. When the reaction time is up, turn off the heat, let the setup cool, dismantle it, and collect and purify the product.

Open end

Water outlet

Cold water inlet

Condensing solvent

Clamp

Heat source

Flask with solvent
and reactants

FIGURE 22.1 A reflux setup.

A DRY REFLUX

If you have to keep the atmospheric water vapor out of your reaction, you must use a
drying tube and the **inlet adapter** in the reflux setup (Fig. 22.2). You can use these
if you need to keep water vapor out of any system, not just the reflux setup.

1. If necessary, *clean and dry the drying tube.* You don't have to do a thorough
   cleaning unless you suspect that the anhydrous drying agent is *no longer anhydrous.* If the stuff is caked inside the tube, it is probably dead. You should
   clean and recharge the tube at the beginning of the semester. Be sure to use
   anhydrous calcium chloride or sulfate. It should last one semester. If you are
   fortunate, **indicating Drierite,** a specially prepared anhydrous calcium sulfate, might be mixed in with the white Drierite. If the color is *blue,* the drying
   agent is good; if it's *red,* the drying agent is no longer dry, and you should get
   rid of it (see Chapter 10, "Drying Agents").

2. Put in a loose plug of *cotton* to keep the drying agent from falling into the reaction flask.

Open end

Drying tube

Drying agent

Cotton

Inlet adapter

Water outlet

Condensing solvent

Cold water inlet

Clamp

Flask with solvent
and reactants

Heat source

FIGURE 22.2 Reflux setup à la
drying tube.

3. Assemble the apparatus as shown in Fig. 22.2, with the *drying tube and adapter on top of the condenser.*

4. At this point, reagents may be added to the flask and heated with the apparatus. Usually, the apparatus is heated while empty to drive water off its walls.

5. Heat the apparatus, usually empty, on a steam bath, giving the entire setup a quarter-turn every so often to heat it evenly. A burner can be used *if there is no danger of fire* and if heating is done carefully. The heavy ground-glass joints will crack if they are heated too much.

6. Let the apparatus cool to room temperature. As it cools, air is drawn through the drying tube before it hits the apparatus. The moisture in the air is trapped by the drying agent.

7. Quickly add the dry reagents or solvents to the reaction flask, and reassemble the system.

8. Carry out the reaction as usual, like a standard reflux.

ADDITION AND REFLUX

Every so often you have to add a compound to a setup while the reaction is going on, usually along with a **reflux.** Well, in adding new reagents, you *don't break open the system, let toxic fumes out, and make yourself sick.* You use an **addition funnel.** Now, we talked about addition funnels back with **separatory funnels** (Chapter 15) when we were considering the **stem,** and that might have been confusing.

Funnel Fun

Look at Figure 22.3*a*. It is a true sep funnel. You put liquids in here and shake and extract them. But could you use this funnel to add material to a setup? *NO.* No ground-glass joint on the end, and only glass joints fit glass joints. Right? Of course, right.

Figure 22.3*c* shows a **pressure-equalizing addition funnel.** See that side arm? Remember when you were warned to remove the stopper of a separatory funnel so that you wouldn't build up a vacuum inside the funnel as you emptied it? Anyway, the side arm equalizes the pressure on both sides of the liquid you're adding to the flask, so it'll flow freely without vacuum buildup and without any need for you to remove the stopper. This equipment is very nice, very expensive, very limited, and very rare. And if you try an **extraction** in one of these, all the liquid will bypass the stopcock and run out the tube onto the floor as you shake the funnel.

So a compromise was reached (Fig. 22.3*b*). Since you'll probably do more extractions than additions, with or without reflux, the pressure-equalizing tube is gone, but the ground-glass joint stayed on. Extractions? No problem. The nature of the stem is unimportant. But during additions, you'll have to take the responsibility to see that nasty vacuum buildup doesn't occur. You can remove the stopper every

Pressure-equalizing tube

Plain stem

(a)

Ground-glass stem with drip tip

(b)

(c)

FIGURE 22.3 Separatory funnels in triplicate. (a) Plain. (b) Compromise separatory addition funnel. (c) Pressure-equalizing addition funnel.

so often, or put a **drying tube** and an **inlet adapter** in place of the stopper. The inlet adapter keeps moisture out and prevents vacuum buildup inside the funnel.

How to Set Up

There are at least two ways to set up an addition and reflux, using either a **three-neck flask** or a **Claisen adapter.** I thought I'd show both of these setups with **drying tubes.** They keep the moisture in the air from getting into your reaction. If you don't need them, do without them.

Often the question comes up, "If I'm refluxing one chemical, how fast can I add the other reactant?" Try to follow your instructor's suggestions. Anyway, usually the reaction times are fixed. So I'll tell you what NOT, repeat NOT, to do.

FIGURE 22.4 Reflux and addition by Claisen tube.

If you reflux something, there should be a little ring of condensate, sort of a cloudy, wavy area in the barrel of the reflux condenser (Figs. 22.4 and 22.5). Assuming an exothermic reaction—the usual case—adding material from the funnel has the effect of heating up the flask. The ring of condensate begins to move *up*. Well, don't *ever* let this get more than three-quarters up the condenser barrel. If the reaction is that fast, a very little extra reagent or heating will push that ring out of the condenser and possibly into the room air. No, no, no, no.

Charged drying tube

Water outlet

Open stopper from time to time to break the vacuum buildup as liquid is added!

Condensing solvent

Flask with solvent and reactants

Heat source

FIGURE 22.5 Reflux and addition by three-neck flask.

EXERCISE

If one of the reagents in a reaction you are using a reflux to carry out is not entirely soluble in the boiling solvent, can you still reflux? *Should* you still reflux? What considerations would come into play here?

REFLUX: microscale

There's not much difference between the microscale reflux and the bigger ones other than the size of the glassware. So, much of the information for the larger setup will still apply (see Chapter 22, "Reflux and Addition"). And if you want a dry microscale reflux, then add a drying tube (Fig. 23.1).

ADDITION AND REFLUX: MICROSCALE

Once you get used to the idea of handling reagents in syringes (see Chapter 8, "Syringes, Needles, and Septa"), addition on a microscale is pretty straightforward (Fig. 23.2). Sometimes you have to watch out for back-pressure in the setup, and you have to hold the syringe plunger a bit to keep it from blowing back out. Normally, you just hold the barrel of the loaded syringe, pierce the septum, and then add the reagent, a little at a time, into the setup. How fast? Usually, as fast as is practical. If the reaction speeds up too much, the ring of condensate in the reflux condenser will travel up and out of the condenser and into the room air. No, no, no, no. This sounds identical to the warning I give for the large-scale setup. *Hint:* It is.

FIGURE 23.1 Microscale reflux, both wet and dry.

Syringe

Septum
screwcap

Clamp

Drying agent

CaCl₂

Cotton plug

Claisen
adapter

Air
condenser

Magnetic
spin vane

Hot plate/magnetic stirrer

Temp. control

Stir control

FIGURE 23.2 A dry microscale reflux
and addition.

SUBLIMATION

■ *You must use a water trap.*

Sublimation occurs when you heat a solid and it turns directly into a vapor. It does not pass GO, nor does it turn into a liquid. If you reverse the process—cool the vapor so that it turns back into a solid—you've *condensed the vapor*. Use the unique word **sublime** for the direct conversion of *solid to vapor*. **Condense** can refer to either *vapor-to-solid* or *vapor-to-liquid* conversions.

Figure 24.1 shows three forms of sublimation apparatus. Note all the similarities. Cold water goes in and down into a **cold finger** on which the vapors from the crude solid condense. The differences are that one is larger and has a **ground-glass joint.** The **side-arm test tube** with **cold-finger condenser** is much smaller. To use them,

1. Put the crude solid into the bottom of the sublimator. How much crude solid? This is rather tricky. You certainly don't want to start with so much that it touches the cold finger. And since, as the purified solid condenses on the cold finger, it begins to grow down to touch the crude solid, there has to be quite a bit of room. I suggest that you see your instructor, who may want only a small amount purified.

FIGURE 24.1 King-size, miniature, and inexpensive microscale sublimation apparatus.

2. Put the cold finger into the bottom of the sublimator. Don't let the clean cold finger touch the crude solid. If you have the sublimator with the ground-glass joint, lightly (and I mean lightly) grease the joint. Remember that greased glass joints should NOT be clear all the way down the joint.

3. Attach the hoses. *Cold water goes in the center tube,* pushing the warmer water out the side tube. Start the cooling water. Be careful! The inexpensive microscale sublimator needs no running cooling water—only ice and water in the tube.

4. If you're going to pull a vacuum in the sublimator, do it now. If the vacuum source is a **water aspirator,** *put a water trap between the aspirator and the sublimator.* Otherwise, you may get depressed if, during a sudden pressure drop, water backs up and fills your sublimator. Also, start the vacuum *slowly.* If not, air, entrained in your solid, comes rushing out and blows the crude product all over the sublimator, like popcorn.

5. When everything has settled down, slowly begin to heat the bottom of the sublimator, if necessary. You might see vapors coming off the solid. Eventually, you'll see crystals of *purified solid form on the cold finger.* Since you'll work with different substances, different methods of heating will have to be used. Ask your instructor.

6. Now the tricky parts. You've let the sublimator cool. If you have a vacuum in the sublimator, carefully—very carefully—introduce air into the device. A sudden inrush of air, and PLOP! Your purified crystals are just so much yesterday's leftovers. Start again.

7. Now again, carefully—very carefully—remove the cold finger, with your pristine product clinging tenuously to the smooth glass surface, without a lot of bonking and shaking. Otherwise, PLOP! et cetera, et cetera, et cetera. Clean up and start again.

MICROSCALE BOILING POINT

The boiling point of a liquid is the temperature at which the vapor pressure of the liquid equals the atmospheric pressure. More practically, you see the liquid boil. The absolute best way to get the boiling point of a liquid is to distill it, and, unfortunately, as the bean counters exert more and more influence over the organic laboratory, there's less and less of a chance you'll ever get to determine a boiling point this way. If you're lucky enough to have this luxury, then go back and read the appropriate sections (see Chapter 19, "Distillation," and Chapter 20, "Microscale Distillation," as appropriate). The rest of you poor devils will be using a technique developed years ago to determine the boiling point of a liquid on a scale of 2 to 5 traditional eyedropper drops.

I didn't have to distinguish drop size before the introduction of Pasteur pipets into the laboratory, but I do now. The drop size of a Pasteur pipet is very different from that of a drop from a traditional eyedropper. I took the average mass of 20 drops of room-temperature cyclohexanone and found that an eyedropper delivered about 0.0201 g and a Pasteur pipet delivered about 0.0137 g. In the following discussions, I'm using eyedropper drops, OK?

MICROSCALE BOILING POINT

Back in the old days, you would take a small (75×150-mm) test tube, strap it to a thermometer with a rubber band or ring, and put about 4–5 drops of your unknown liquid into the test tube. Then you'd get an ordinary melting-point capillary tube sealed on one end, drop the melting-point tube into the liquid in the test tube *open end down,* and clamp the whole shebang, so that the thermometer bulb, the test tube with your sample, and the open-end-down melting-point capillary in it were all in an oil bath (Fig. 25.1).

At this point you'd be told to heat the bath and watch the open end of the capillary tube. As you heated your liquid unknown, first the air in the capillary would expand, and a very slow stream of bubbles would start to come out of the tube as the air left it. At some point—and I was never really sure about this point myself—the stream of bubbles was to get faster as the liquid started to boil. You'd let this go for

Thermometer

Small test tube
Sealed end up
Rubber band or ring
Upside-down sealed-end capillary tube
Liquid unknown
Open end down— bubbles come out here

Oil bath

FIGURE 25.1 Micro boiling-point apparatus.

awhile and then remove the heat. Your boiling liquid would cool, and as soon as the liquid entered the capillary tube—BINGO!—you'd read the thermometer and get the first value of the boiling range. When the liquid finally filled the tube, you'd read the temperature again and get the second value of the boiling range. There were, of course, a few things to watch out for:

1. **When do the air bubbles become sample vapor bubbles?** Good question, and the one that still bothers me the most. Ideally, you won't rush heating the sample, and there will be a noticeable difference between the rate of formation of air expulsion bubbles and actual boiling bubbles. If your sample boils at a relatively high temperature (probably >100°C), you have more time to heat it and convince yourself that the bubbles you see are those of your boiling liquid and not just air trying to leave the tube. Try this with, say, acetone (bp 56°C), and you could boil away all of the liquid faster than you can decide if the liquid is boiling away.

    If you're sure your liquid doesn't decompose, react, or change in some way when boiling, you can stop the heating after what you believe to be the rapid-boiling bubbles appear. When that sample cools and the liquid rushes back into the tube and fills it completely, you know those bubbles were not air, but boiling sample. The advantage you have now is that when you reheat the sample, any bubbles you see can't be air bubbles; they have to be boiling sample bubbles.

    Now when you reheat the sample and the stream of bubbles starts again, you can be confident that the bubbles are not air, but your boiling sample. Let the sample cool, and the temperatures you record are the boiling range, called boiling point, of your liquid sample. Again, the disadvantage is that the sample may decompose or in some way change when boiling, and your data are meaningless.

2. **I only got one reading on my thermometer.** A melting point can appear to be one temperature, rather than a range, however small, if you heat the sample too rapidly. So also for this method of boiling-point determination, you can appear to get one temperature, rather than a range, however small, if you cool the sample too rapidly.

ULTRAMICROSCALE BOILING POINT

The only modification, if you can call it that, in the ultramicroscale technique is to perform it on an even smaller scale. The test tube into which you put your sample becomes a melting-point capillary, and the closed-end-up, open-end-down melting-point capillary is a smaller version you have to make yourself! Out of another melting-point capillary, no less.

Heat the midsection of an open-end melting-point capillary tube, and carefully draw the glass to form an even smaller capillary section. Break this section in two—just as if you were making a capillary spotter (see Chapter 27, "Thin-Layer Chromatography"). At this point you seal the narrow end on each "capillary spotter" and then break this narrow tube off the original, unchanged larger section of the melting-point capillary tube. Violà—two itty-bitty, really narrow capillary tubes. Put these tubes, which are about 5 to 8 mm long, into the bigger melting-point capillary, open end down, closed end up. If this seems familiar, well, it should. It's a smaller version of the test-tube-and-melting-point-capillary setup described earlier. And you heat this by putting it in a melting-point apparatus! (See Chapter 12, "The Melting-Point Experiment.") As usual, in addition to the same problems of determining the boiling point that you've seen for the larger setup above, this one has its own set of problems:

1. *Loading the melting-point capillary with sample.* The only practical way to do this is with a syringe. You load a syringe with your liquid and squirt it into the capillary tube (see Chapter 8, "Syringes, Needles, and Septa"). The needles are usually too short to place the liquid directly at the bottom of the melting-point capillary, so you'll probably have to centrifuge the tube to force the liquid to the bottom.

2. *Making the itty-bitty, really narrow capillary not narrow enough.* If this is the case, then your liquid sample will wick up between the walls of the melting-point capillary and your handmade capillary, and you might not have a big enough pool of liquid at the bottom of the melting-point capillary for those bubbles to form.

3. *Making the itty-bitty, really narrow capillary not heavy enough.* If your handmade capillary is too light (usually from being too short), then it can float around in the melting-point capillary and give results that might be difficult to interpret. One way of making sure that your homemade capillary is heavy enough (besides making it overly long—don't do this! Anything about 5 to 8 mm long is just about right!) is to make sure you leave a bit of fused glass on the end to help weigh the homemade capillary down. Be careful, though. If you get too creative and leave a really unusual glob of glass on the end, your homemade capillary might get stuck at the top of the melting-point capillary.

Well, that's about the end of the ultramicroscale boiling-point technique, but for the fact that you still have the same problems as with the microscale boiling-point technique, already described.

CHROMATOGRAPHY: some generalities

Chromatography is the most useful means of separating compounds to purify and identify them. Indeed, separations of colored compounds on paper strips gave the technique its colorful name. Although there are many different types of chromatography, all the forms bear tremendously striking similarities. **Thin-layer, wet-column,** and **dry-column chromatography** are common techniques that you'll run across.

This chromatography works by differences in **polarity.** (That's not strictly true for all types of chromatography, but I don't have the inclination to do a 350-page dissertation on the subject, when all you might need to do is separate the differently colored inks in a black marker pen.)

ADSORBENTS

The first thing you need is an **adsorbent,** a porous material that can suck up liquids and solutions. Paper, silica gel, alumina (ultrafine aluminum oxide), cornstarch, and kitty litter (unused) are all fine adsorbents. Only the first *three* are used for chromatography, however. You may or may not need a **solid support** with these. Paper hangs together, is fairly stiff, and can stand up by itself. Silica gel, alumina, cornstarch, and kitty litter are more or less powders and need a solid support such as a glass plate to hold them.

Now you have an *adsorbent on some support*, or a *self-supporting adsorbent*, such as a strip of paper. You also have a mixture of stuff you want to separate. So you dissolve the mixture in an easily evaporated solvent, such as methylene chloride, and put some of it on the adsorbent. *Zap!* It is adsorbed! Stuck on and held to the adsorbent. But because you have a mixture of different things and they are different, they will be *held on the adsorbent in differing degrees.*

SEPARATION OR DEVELOPMENT

Well, so there's this mixture, sitting on this adsorbent, looking at you. Now you start to *run solvents through the adsorbent*. Study the following list of solvents. Chromatographers call these solvents **eluents** (or **eluants**).

THE ELUATROPIC SERIES

Not at all like the World Series, **the eluatropic series** is simply a list of solvents arranged according to increasing polarity.

*Some solvents arranged
in order of increasing polarity*

(Least polar)	Petroleum ether
	Cyclohexane
Increasing	Toluene
polarity	Chloroform
	Acetone
	Ethanol
(Most polar)	Methanol

So you start running "petroleum ether" (remember, pet. ether is a mixture of hydrocarbons like gasoline—it's not a true ether at all). It's not very polar. So it is *not held strongly to the adsorbent.*

Well, this solvent is traveling through the adsorbent, minding its own business, when it encounters the mixture placed there earlier. It tries to kick the mixture out of the way. But most of *the mixture is more polar, held more strongly* on the adsorbent. Since the pet. ether cannot kick out the compounds that are more polar than itself very well, most of *the mixture is left right where you put it.*

No separation.

Desperate, you try methanol, one of the most polar solvents. It is *really held strongly to the adsorbent.* So it comes along and kicks the living daylights out of just about *all* of the molecules in the mixture. After all, the methyl alcohol *is more polar, so it can move right along and displace the other molecules.* And it does. So, when you evaporate the methanol and look, *all the mixture has moved with the methanol,* so you get *one spot* that moved right with the **solvent front.**

No separation.

Taking a more reasonable stand, you try chloroform, because it has an intermediate polarity. The chloroform comes along, sees the mixture, and is able to push out, say, *all but one* of the components. As it travels, kicking the rest along, it gets tired and starts to leave some of the more polar components behind. After a while, *only one component is left moving with the chloroform,* and that may be dropped, too. So, at the end, *there are several spots left*, and each of them is in a *different place* from where it was at the start. Each spot is *at least one different component* of the entire mixture.

Separation. At last!

You could also use a solvent mixture if none of these pure solvents gives you the separation you want. Unfortunately, the effects don't just add up: A little bit of a polar solvent goes a long way. A 50–50 mixture of any two solvents is going to wind up with a polarity closer to the more polar solvent, *not* in the middle at all.

I picked these solvents for illustration. They are quite commonly used in this technique. I worry about the hazards of using chloroform, however, because it's been implicated in certain cancers. Many other common solvents are also suspected carcinogens. In the lab, either you will be told what solvent (eluent) to use or you will have to find out yourself, mostly by trial and error.

Thin-layer chromatography (TLC) is used for identifying compounds and determining their purity. You can even follow the course of a reaction by using the technique. The most common adsorbent used is **silica gel,** but **alumina** is gaining popularity, for good reason. Compounds should separate the same on an alumina plate as on an alumina column, and column chromatography using alumina is still very popular. It is very easy to run test separations on **TLC plates** rather than on **chromatographic columns.**

Nonetheless, both of these adsorbents are powdered and require a **solid support. Microscope slides** are extremely convenient. To keep the powder from just falling off the slides, manufacturers add a *gypsum binder* (plaster). Adsorbents with the binder usually have a "G" stuck on the name or say "For thin-layer use" on the container.

Sometimes a fluorescent powder is put into the adsorbent to help with **visualization** later. The powder usually glows a bright green when you expose it to **254-nm-wavelength ultraviolet (UV) light.** You can probably figure out that if a container of silica gel is labeled **Silica Gel G-254,** you've got a TLC adsorbent with all the bells and whistles.

Briefly, you mix the adsorbent with water, spread the mix on the microscope slide in a *thin layer,* let it dry, and then activate the coating by heating the coated slide on a hot plate. After the plate cools, you **spot** or place your unknown compound on the plate, let an **eluent** run through the adsorbent (**development**), and finally examine the plate (**visualization**).

PREPARATION OF TLC PLATES

1. Clean and dry several microscope slides.
2. Weigh out some adsorbent in an Erlenmeyer flask, and add water.
   a. For silica gel, use a 1 : 2 ratio of gel to water. About 2.5 g gel and 5 g water will do for a start.
   b. For alumina, use a 1 : 1 ratio of alumina to water. About 2.5 g alumina and 2.5 g water will be a good start.
3. Stopper the flask and shake it until all the powder is wet. This material must be used quickly because a gypsum (plaster) binder is present.
4. Spread the mix by using a *medicine dropper. Do not use disposable pipets!* The disposable pipets have *extremely* narrow openings at the end, and they clog up easily. A "dipping method" exists for preparing TLC slides, but since the usual solvents, methanol and chloroform (**Caution!** Toxic!), do not activate the binder, the powder falls off the plate. Because the layers formed by this process are very thin, they are very fragile.
5. *Run a bead of mix around the outside of the slide, and then fill the remaining clear space. Leave $\frac{1}{4}$ in. of the slide blank on one end,* so you can hold onto it. *Immediately* tap the slide from the bottom to smooth out the mix (Fig. 27.1).

223

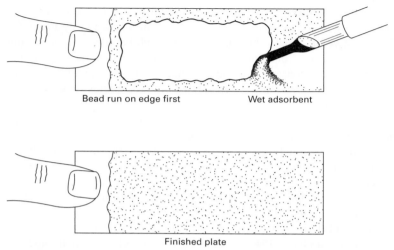

Bead run on edge first   Wet adsorbent

Finished plate

FIGURE 27.1 Spreading adsorbent on a TLC plate.

Repeat this procedure with as many slides as you can. If the mix sets up and becomes unmanageable, add a little water and shake well.

6. Let the slides sit until the gloss of water on the surface has gone. Then place the slides on a hot plate until they dry.

CAUTION! If the hot plate is too hot, the water will quickly turn to steam and blow the adsorbent off the slides.

Pre-prepared TLC Plates

There are **pre-prepared plates** that have an active coating on a thin plastic sheet, also with or without the fluorescent indicator. You can cut these to any size (they are about 8 by 8 in.) with a pair of scissors. *Don't touch the active surface with your fingers—handle them only by the edge.* The layers on the plate are much thinner than those you would make by spreading adsorbent on a microscope slide, so you have to use smaller amounts of your compounds in order not to overload the adsorbent.

THE PLATE SPOTTER

1. The **spotter** is the apparatus used to put the solutions that you want to analyze on the plate. You use it to make a spot of sample on the plate.

2. Put the center of a **melting-point capillary** into a small, blue Bunsen burner flame. Hold it there until the tube softens and starts to sag. Do not rotate the tube, ever.

FIGURE 27.2 Making capillary spotters from melting-point tubes.

3. *Quickly* remove the capillary from the flame, and pull both ends (Fig. 27.2). If you leave the capillary in the flame too long, you get an obscene-looking mess.

4. Break the capillary at the places shown in Figure 27.2 to get **two spotters** that look roughly alike. (If you've used capillaries with *both ends open already,* then you don't have a closed end to break off.)

5. Make up 20 of these, or more. You'll need them.

6. Because TLC is so sensitive, spotters tend to "remember" old samples if you reuse them. *Don't put different samples in the same spotter.*

SPOTTING THE PLATES

1. You might want to lightly mark the places where you're going to spot a plate with a pencil (never ink! *Never!*), especially if your sample is dilute and you have to spot more than once in the sample place.

2. Dissolve a small portion (1–3 mg) of the substance you want to chromatograph in *any* solvent that dissolves it *and* evaporates rapidly. Dichloromethane or diethyl ether often works best.

3. Put the thin end of the capillary spotter into the solution. The solution should rise up into the capillary.

4. Touch the capillary to the plate *briefly!* The compound will run out and form a small spot. Try to keep the spot as small as possible—not larger than $\frac{1}{4}$ in. in diameter. Blow gently on the spot to evaporate the solvent. Touch the capillary to the *same place.* Let this new spot grow to be *almost the same size as the one already there.* Remove the capillary and evaporate the solvent by gently blowing on the spot. This will build up a concentration of the compound.

FIGURE 27.3 Putting a spot of compound on a TLC plate.

5. Take a sharp object (an old pen point, capillary tube, spatula edge, etc.) and *draw a straight line through the adsorbent,* as close to the clear glass end as possible (Fig. 27.3). Make sure that it runs all the way across the end of the slide and goes right down to the glass. This will keep the solvent from running up to the ragged edge of the adsorbent. It will travel only as far as the smooth line you have drawn. Measurements will be taken from this line.

6. Now make a small notch on the plate coating at the level of the spots to mark their starting position. You'll need this later for measurements.

Pre-prepared plates have a very thin layer of adsorbent and a clear plastic backing, making them translucent to a degree. You can draw a template of a TLC plate in your notebook with a line near the bottom, and, with the actual TLC plate on top of it, spot the plate using the image of the line that you can just see through the plate itself. Later, after the plate develops, you can put it back on this template and use the line as the basis for your measurements (Fig. 27.4).

DEVELOPING A PLATE

1. Take a 150-mL beaker, line the sides of it with filter paper, and cover with a watch glass (Fig. 27.5). Yes, you can use other containers. I've used beakers because all chemistry stockrooms have beakers, you know what beakers look like, and it's a handy single word for the more cumbersome "thin-layer development chamber used at your institution." So if you do use jelly jars, drinking glasses, wide-mouth screwcap bottles, or whatever, just substitute that description for beaker in this discussion, OK?

2. Choose a solvent to **develop** the plate. You let this solvent (eluent) pass through the adsorbent by capillary action. Nonpolar eluents (solvents) force nonpolar compounds to the top of the plate, whereas polar eluents force *both polar and nonpolar materials up the plate.* There is only one way to choose eluents: educated guesswork. Use the chart of eluents presented in Chapter 26.

3. Pour some of the eluent (solvent) into the beaker, and tilt the beaker so that the solvent wets the filter paper. Put no more than $\frac{1}{4}$ in. of eluent in the bottom of

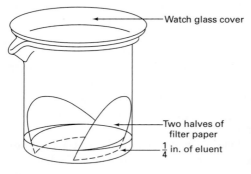

FIGURE 27.4 Putting a spot of compound on a plate using a drawn template underneath.

the beaker! This saturates the air in the beaker with eluent (solvent) and stops the evaporation of eluent from the plate.

4. Place the slide in the developing chamber as shown (Fig. 27.6). *Don't let the solvent in the beaker touch the spot on the plate,* or the spot will dissolve away into the solvent! If this happens, you'll need a new plate, and you'll have to clean the developing chamber as well.

FIGURE 27.5 The secret identity of a 150-mL beaker as a TLC slide development chamber is exposed. You can substitute a jar with a cap.

Watch glass cover

Clear end of plate

Spotted material must be above solvent level

$\frac{1}{4}$ in. of eluent

FIGURE 27.6   Prepared, spotted TLC plate in a prepared developing chamber. You can substitute a jar with a cap.

5. Cover the beaker with a watch glass. The solvent (eluent) will travel up the plate. The filter paper keeps the air in the beaker saturated with solvent, so that it doesn't evaporate from the plate. When the solvent reaches the line, *immediately* remove the plate. Drain the solvent from it, and blow gently on the plate until *all* of the solvent is gone. If not, there will be some trouble visualizing the spots.

Don't breathe in the fumes of the eluents! Make sure you have adequate ventilation. Work in a hood if possible.

VISUALIZATION

Unless the compound is colored, the plate will be blank, and you won't be able to see anything, so *you must visualize the plate.*

1. **Destructive visualization.** Spray the plate with sulfuric acid, then bake in an oven at 110°C for 15–20 minutes. Any spots of compound will be charred blots, utterly destroyed. *All* spots of compound will be shown.

2. **Nondestructive visualization.**
   a. **Long-wave UV (Hazard!).** Most TLC adsorbents contain a fluorescent powder that glows bright green when you put them under long-wave UV light. There are two ways to see the spots:
   (1) The background glows green, and the spots are dark.
   (2) The background glows green, and the spots glow some other color. The presence of *excess eluent may cause whole sections of the plate to remain dark.* Let all of the eluent evaporate from the plate.
   b. **Short-wave UV (Hazard!).** The plates stay dark. *Only the compounds may glow.* This is usually at 180 nm.
      Both of the UV tests can be done in a matter of seconds in a **UV light box** (Fig. 27.7). Since most compounds are unchanged by exposure to UV light, the test is considered **nondestructive.** Not everything will show up, but the procedure

Buttons select short-wave (180-nm), long-wave (250-nm), or visible light

Short- and long-wave selection buttons

Spots and/or fluorescent powder glows

Lift rubber flaps to move plates in and out

Visible-ultraviolet light box

Hand-held UV light

FIGURE 27.7   Visible-UV light box and hand-held UV unit.

is good enough for most compounds. When using the light box, *always turn it off when you leave it.* If you don't, not only does the UV filter burn out, but also your instructor becomes displeased.

Since neither the UV nor the iodine test is permanent, it helps to have a record of what you've seen. You must *draw an accurate picture of the plate in your notebook.* Using a sharp-pointed object (pen point, capillary tube, etc.), you can trace the outline of the spots on the plate while they are under the UV light (**CAUTION!** Wear gloves!) or before the iodine fades from the plate.

3. *Semidestructive visualization.* Set up a developing tank (150-mL beaker) but leave out the filter paper and any solvent. Just a beaker with a cover. Add a few crystals of **iodine.** Iodine vapors will be absorbed onto most spots of compound, coloring them. Removing the plate from the chamber causes the iodine to evaporate from the plate, and *the spots will slowly disappear. Not all spots may be visible.* So if there's nothing there, that doesn't mean nothing's there. The iodine might have reacted with some spots, changing their composition. Hence the name **semidestructive visualization.**

INTERPRETATION

After visualization, there will be a spot or spots on the plate. Here is what you do when you look at them.

1. Measure the distance *from that solvent line* drawn across the plate *to where the spot started.*

2. Measure the distance from where the *spot stopped* to where the spot began. Measure to the *center of the spot* rather than to one edge. If you have more than one spot, get a distance for each. If the spots have funny shapes, do your best.

3. Divide the *distance the solvent moved* into the *distance the spot(s) moved*. The resulting ratio is called the R_f **value.** *Mathematically,* the ratio for any spot should be between 0.0 and 1.0, or you goofed. *Practically,* spots with R_f values greater than about 0.8 and less than about 0.2 are hard to interpret. They could be single spots or multiple spots all bunched up and hiding behind one another.

4. Check out the R_f **value**—it may be helpful. In identical circumstances, this value would always be the *same* for a single compound. If this were true, you could identify unknowns by running a plate and looking up the R_f value. Unfortunately, the technique is not that good, but you can use it *with some judgment* and a *reference compound* to identify unknowns (see "Multiple Spotting," following).

Figures 27.8 to 27.10 provide some illustrations. Look at Figure 27.8: If you had a mixture of compounds, you could never tell. This R_f value gives no information. Run this compound again. Run a new plate. *Never redevelop an old plate!*

Use a more polar solvent!

No information is in Figure 27.9, either. You couldn't see a mixture if it were here. Run a new plate. *Never redevelop an old plate!*

Use a less polar solvent!

If the spot moves somewhere between the two limits (as shown in Fig. 27.10) and *remains a single spot*, the compound is *probably* pure. You should try a second solvent system, or a second adsorbent, before drawing this conclusion. If *more than*

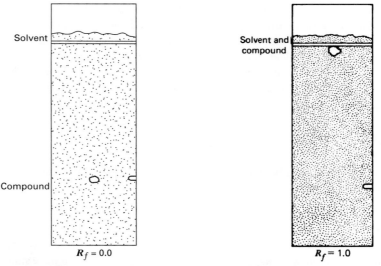

Solvent

Compound

$R_f = 0.0$

FIGURE 27.8 Development with a nonpolar solvent and no usable results.

Solvent and compound

$R_f = 1.0$

FIGURE 27.9 Development with a very polar solvent and no usable results.

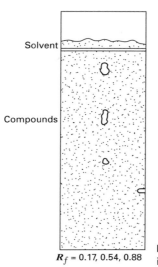

R_f = 0.17, 0.54, 0.88

FIGURE 27.10   Development with just the right solvent is a success.

one spot shows, the compound is definitely impure, and it is a mixture. Whether the compound should be purified is a matter of judgment.

MULTIPLE SPOTTING

You can run more than one spot, either to save time or to make comparisons. You can even *identify unknowns.*

Let's say that there are two unknowns, A and B. Say one of them can be biphenyl (a colorless compound that smells like mothballs). You spot two plates, one with A and biphenyl side by side, and the other with B and biphenyl side by side. After you develop both plates, you have the results shown in Fig. 27.11.

Apparently, A is biphenyl.

Note that the R_f values are not perfect. This is an imperfect world, so don't panic over a slight difference.

Thus, we have a method that can quickly determine:

1. Whether a compound is a mixture.

2. The identity of a compound *if a standard is available.*

COSPOTTING

You can nail down identification with **cospotting,** but it is a bit tricky. This time, spot one plate with your unknown, A, in two places. Now let the spots dry entirely. I mean entirely! Now we'll spot biphenyl *right on top of one of the spots of A.* (If you didn't let the earlier spot dry entirely, the biphenyl spot would bleed and you'd have a *very*

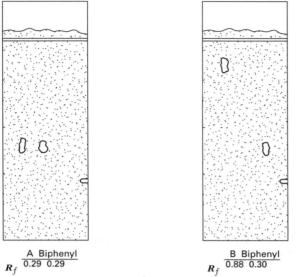

R_f $\dfrac{\text{A Biphenyl}}{0.29 \quad 0.29}$ R_f $\dfrac{\text{B Biphenyl}}{0.88 \quad 0.30}$

FIGURE 27.11 Side-by-side comparison of an unknown and a leading brand known.

	A over biphenyl	B over biphenyl
R_f	0.29	0.88 and 0.29

FIGURE 27.12 Cospotting biphenyl over A and B.

large—and useless—spot of mixed compounds.) Do the same for B on another plate. Now run these plates.

The R_f values might be a bit different, but there's only *one spot* in the biphenyl over the A column. No separation. They are the same. If you see two (or more) spots come from cospotting, the substances, like B, are not the same (Fig. 27.12).

OTHER TLC PROBLEMS

What you've seen is what you get if everything works OK. You can:

1. Make the spots too concentrated. Here spots smear out, and you don't see any separation. Dilute the spotting solution, or don't put so much on (Fig. 27.13).

2. Put the spots too close together. Here you can get the spots bleeding into each other, and you might not be able to tell which spot came from which origin (Fig. 27.14).

3. Spot too close to the edge. Here you get inaccurate R_f values, bleeding, and other problems. The spots are not surrounded as much by adsorbent and solvent, so unequal forces are at work here (Fig. 27.15).

PREPARATIVE TLC

When you use an **analytical technique** (like TLC) and you expect to **isolate compounds,** it's often called a **preparative (prep) technique.** So TLC becomes **prep TLC.** You use the same methods, only on a larger scale.

Concentrated   Diluted
mixture   mixture   **FIGURE 27.13**   Effect of concentration on a separation.

A close to B

FIGURE 27.14 Effect of spots being too close together.

Instead of a microscope slide, you usually use a 12 × 12-in. glass plate and coat it with a thick layer of adsorbent (0.5–2.0 mm). Years ago, I used a small paintbrush to put a line of dissolved sample (a streak rather than a spot) across the plate near the bottom. Now you can get special plate streakers that give a finer line and less spreading. You put the plate in a large developing chamber, develop, and visualize the plate as usual.

The thin line separates and spreads into bands of compounds, much like a tiny spot separates and spreads on the analytical TLC plates. Rather than just look at the bands, though, you *scrape the adsorbent holding the different bands into different flasks*, blast your compounds off the adsorbents with appropriate solvents, filter off the adsorbent, and finally evaporate the solvents and actually recover the separate compounds.

FIGURE 27.15 Effect of spot too close to the edge.

EXERCISES

1. What's the matter? Ballpoint pen not good enough to mark a TLC plate?

2. A TLC plate has a shiny side and a dull side (as if it were coated with some kind of powder). Which should you use to spot your samples on?

3. What are the ways to visualize a TLC plate? Compare, contrast, and all that.

4. Your TLC plate has just finished its run. Why could the entire plate be dark at both wavelengths in the UV light box?

5. Define R_f value.

6. Compounds on TLC plates separate by molecular weight, don't they? Explain.

7. Two spots on a TLC plate have slightly different R_f values. Could they still be the same compound? Explain. How would you spot a new plate to remove all doubt?

8. Describe the effects of the following conditions: Samples are spotted too low on the TLC plate; eluent is too polar; eluent is not polar enough; sample is too concentrated; the top is left off the TLC chamber; the plate is left in the chamber after the solvent has reached the top of the plate.

9. You know there are organic acids on your TLC plates and they are tailing badly. Causes and cures? Is there really any difference in the characteristics of acid-washed alumina and base-washed silica gel?

Wet-column chromatography, as you may have guessed, is chromatography carried out on a **column of adsorbent** rather than a layer. It is cheap, easy, and carried out at room temperature, and you can also separate large amounts, *gram quantities,* of mixtures.

In column chromatography, the adsorbent is either alumina or silica gel. Alumina is basic and silica gel is acidic. If you try out an eluent (solvent) on silica gel plates, you should use a silica gel adsorbent. And if you have good results on alumina TLC, use an alumina column.

Now you have a *glass tube as the support* that holds the adsorbent in place. You dissolve your mixture and put it on the adsorbent at the top of the column. Then you wash the mixture down the column using at least one eluent (solvent), perhaps more. The compounds carried along by the solvent are washed *entirely out of the column,* into separate flasks. Then you isolate the separate fractions.

PREPARING THE COLUMN

1. The adsorbent is supported by a glass tube that has either a stopcock or a piece of tubing and a screw clamp to control the flow of eluent (Figs. 28.1 and 28.2). You can use an ordinary buret. What you will use will depend on your own lab program. Right above this control you put a wad of cotton or glass wool to keep everything from falling out. *Do not use too much cotton or glass wool,* and *do not pack it too tightly.* If you ram the wool into the tube, the flow of eluent will be very slow, and you'll be in the lab until next Christmas waiting for the eluent. If you pack it *too loosely,* all the stuff in the column *will fall out.*

2. At this point, fill the column halfway with the least polar eluent you will use. If this is not given, you can guess it from a quick check of the separation of the mixture on a TLC plate. This would be the advantage of an alumina TLC plate.

3. Slowly put sand through a funnel into the column until there is a 1-cm layer of sand over the cotton. Adsorbent alumina or silica gel is so fine that either is likely to go through cotton but not through a layer of fine sand.

4. During this entire procedure, *keep the level of the solvent above that of any solid material in the column!*

5. Now slowly add the adsorbent. Alumina and silica gel are adsorbents, and they'll suck up the solvent. When they do, heat is liberated. The solvent may boil and *ruin the column.* Add the adsorbent slowly! Use about 25 g of adsorbent for every 1 g of mixture you want to separate. While adding the adsorbent, *tap or gently swirl the column to dislodge any adsorbent or sand on the sides.* You know that a plastic wash bottle with eluent in it can wash the stuff down the sides of the column very easily.

6. When the alumina or silica gel settles, you normally have to add sand (about 1 cm) to the top to keep the adsorbent from moving around.

Solvent: Entire column must always
be under solvent

Sand (1 cm)

Adsorbent: 25 g for every 1 g
of mixture to
be separated

Sand (1 cm)

Glass wool or cotton

Stopcock to control flow

Collection flask

FIGURE 28.1 Wet-column chromatography setup.

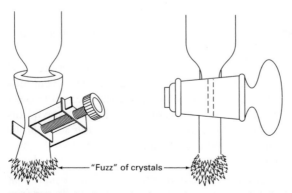

"Fuzz" of crystals

FIGURE 28.2 A growth of crystals occurs as the eluent evaporates.

7. Open the stopcock or clamp, and let solvent out until the level of the solvent is just above the upper level of sand.

8. *Check the column! If there are air bubbles or cracks in the column, dismantle the whole business and start over!*

COMPOUNDS ON THE COLUMN

If you've gotten this far, congratulations! Now you have to get your mixture, the analyte, on the column. Dissolve your mixture in the same solvent that you are going to put through the column. Try to keep the volume of the solution or mixture as *small as possible*. If your mixture does not dissolve entirely (*and it is important that it do so*), check with your instructor! You might be able to use different solvents for the analyte and for the column, but this isn't as good. You might use the *least polar* solvent that will dissolve your compound.

If you must use the column eluent as the solvent and not all of the compound dissolves, you can filter the mixture through filter paper. Try to keep the volume of solution down to 10 mL or so. After this, the sample becomes unmanageable.

1. Use a pipet and rubber bulb to slowly and carefully add the sample to the top of the column (Fig. 28.3). *Do not disturb the sand!*

2. Open the stopcock or clamp, and let solvent flow out until the level of the solution of compound is slightly above the sand. *At no time let the solvent level get below the top layer of sand!* The compound is now "on the column."

Solution of mixture to be separated

Solution run down side, not dropped in

Solvent just above sand
Sand
Alumina

FIGURE 28.3   Putting compounds on the column by pipet.

3. Now add eluent (solvent) to the column above the sand. *Do not disturb the sand!* Open the stopcock or clamp. Slowly let eluent run through the column until the first compound comes out. Collect the different products in Erlenmeyer flasks. You may need lots and lots of Erlenmeyer flasks. *At no time let the level of the solvent get below the top of the sand!* If necessary, stop the flow, add more eluent, and start the flow again.

VISUALIZATION AND COLLECTION

If the compounds are colored, you can watch them travel down the column and separate. If one or all are colorless, you have problems. So,

1. *Occasionally* let one or two drops of eluent fall on a clean glass microscope slide. Evaporate the solvent and see if there is any sign of your crystalline compound! This is an excellent spot test, but don't be confused by nasty plasticizers from the tubing (Fig. 28.2) that try to put one over on you by pretending to be your product.

2. Put the narrow end of a "TLC spotter" to a drop coming off at the column. The drop will rise up into the tube. Using this loaded spotter, *spot, develop, and visualize a TLC plate with it.* Not only is this more sensitive, but you can also see whether the stuff coming out of the column is pure (see Chapter 27, "Thin-Layer Chromatography"). You'll probably have to collect more than one drop on a TLC plate. If it is very dilute, the plate will show nothing, even if there actually is compound there. It is best to sample four or five consecutive drops.

Once the first compound or compounds have come out of the column, those that are left may move down the column much too slowly for practical purposes. Normally you start with a nonpolar solvent. But by the time all the compounds have come off, it may be time to pick up your degree. The solvent may be too nonpolar to kick out later fractions. So you have to decide to change to a more polar solvent. This will kick the compound right out of the column.

To change solvent in the middle of a run:

1. Let the old solvent level run down to just above the top of the sand.

2. *Slowly add new,* **more polar solvent,** *and do not disturb the sand.*

You, and you alone, have to decide if and when to change to a more polar solvent. (Happily, sometimes you'll be told.)

If you have only two components, start with a nonpolar solvent, and when you are sure the *first component is completely off the column, change to a really polar one*. With only two components, it doesn't matter what polarity solvent you use to get the second compound off the column.

Sometimes the solvent evaporates quickly and leaves behind a "fuzz" of crystals around the tip (Fig. 28.2). Just use some fresh solvent to wash them down into the collection flask.

Now all the components are off the column and in different flasks. Evaporate the solvents (*no flames!*), and lo! The crystalline material is left.

Dismantle the column. Clean up. Go home.

WET-COLUMN CHROMATOGRAPHY: MICROSCALE

Occasionally, you'll be asked to clean and/or dry your product using **column chromatography** by pipet. It's a lot like large-scale, wet-column chromatography.

One problem with this pipet column is that there's no stopcock. Once it starts running, it goes until it runs out. And you can't let the column dry out, either. So it looks like you're committed to running the column once you prepare it.

In general (Fig. 28.4),

1. Shorten the tip of a Pasteur pipet (see "Pipet Cutting" in Chapter 7, "Pipet Tips") and tamp cotton into the tip as usual.

2. Add about 50 mg of fine sand.

Drip mixture down
side of pipet

Tiny clamp

Solvent to just
top of sand

More sand

Adsorbent

Fine sand (50 mg)

Cotton

Cut tip

FIGURE 28.4   Pasteur pipet wet-column chromatography.

3. Now comes the adsorbent, about 500 mg or so. Gently tap the pipet to pack this down.

4. Put a bit more sand on top to keep the adsorbent from flying around.

5. Now slowly add the solvent, and let it wet the entire column.

6. Dissolve your sample in a minimum amount of solvent—the least polar solvent in which it'll dissolve—and add this solution slowly to the top of the pipet. Don't fill the pipet with liquid; drip the solution onto the sand.

7. With the sample all on the column, just below the sand add a small amount of **elution solvent.** This may or may not be different from the solvent you dissolved your sample in.

8. Wait and watch. Watch and wait. Do not ever let the column run dry. Collect the drops off the end in a series of fractions in some small containers.

FIGURE 28.5 Professional flash chromatography columns. (*a*) Conventional. (*b*) Screw-thread. From Mayo, Pike, and Tumper, *Microscale Organic Laboratory*, 3rd ed., Wiley, New York, 1994.

FLASH CHROMATOGRAPHY

Take an ordinary wet-column chromatography setup, put it under pressure, and you've got **flash chromatography.** Usually this takes special apparatus (Fig. 28.5), but the hard part, construction of the column and putting the sample on the column, is as described above. There have been a number of modifications to adapt this technique to the existing, well, less expensive equipment of the undergraduate organic laboratory. These range from using a fish aquarium air pump to provide the necessary air pressure (but I do have concerns about solvents, some of which might be very flammable in the proximity of non-laboratory-approved switches and motors) to placing a Y-shaped glass tube in a single-holed rubber stopper, connecting a source of air or N_2 gas to one end, and "regulating" the pressure to the column with a well-placed thumb on the other arm of the Y.

And because the column is under pressure, the caution of packing the column or jamming the cotton or glass wool into the tip too tightly now has a double edge to it: Go ahead! . . . Packing the column too tightly will matter less, as the gas pressure will compensate and push the solvent through anyway. Hold it! . . . Packing the column too tightly with the whole system under gas pressure might be considered foolhardy. You, unfortunately, must be the judge. And don't forget to release the pressure if you decide to stop the flow at the stopcock or pinch clamp, unless you believe the system can hold up under the pressure.

MICROSCALE FLASH CHROMATOGRAPHY

Put a rubber bulb over the Pasteur pipet wet-column chromatography setup (Fig. 28.4) and press it. Don't let go, unless you want the eluent to slurp back up to the bulb and ruin the setup. Remove the bulb, let it go, put it back on the pipet . . . press. Remove the bulb, let it go, put it back on the pipet . . . press. Remove the bulb, let it go, put it back on the pipet . . . press. Remove the bulb . . .

EXERCISES

1. Compare/contrast wet-column chromatography with TLC.
2. If you don't want the column packing coming out of the column, you have to cram the cotton into the tip with lots of force. Right?

When light travels from one medium to another, it changes velocity and direction a bit. If you've ever looked at a spoon in a glass of water, you've seen this: The image of the spoon in water is displaced a bit from the image of the spoon in air, and the spoon looks broken. When the light rays travel from the spoon in the water and break out into the air, they are **refracted,** or shifted (Fig. 29.1). If we take the ratio of the sine of the angles formed when a light ray travels from air to water, we get a single number, the **index of refraction,** or **refractive index.** Because we can measure the index of refraction to a few parts in 10,000, this is a very accurate physical constant for identifying a compound.

The refractive index is usually reported as n_D^{25}, where the tiny 25 is the temperature at which the measurement was taken and the tiny capital D means we've used light from a sodium lamp, specifically a single yellow frequency called the sodium D line. Fortunately, you don't have to use a sodium lamp if you have an Abbé refractometer.

THE ABBÉ REFRACTOMETER (FIG. 29.2)

The refractometer looks a bit like a microscope. It has:

1. *An eyepiece.* You look in here to make your adjustments and read the refractive index.

2. *A compensating prism adjust drum.* Since the Abbé refractometer uses white light and not light of one wavelength (the sodium D line), the white light *disperses* as it goes through the optics, and rainbowlike color fringing shows up when you examine your sample. By turning this control, you rotate some Amici compensation prisms that eliminate this effect.

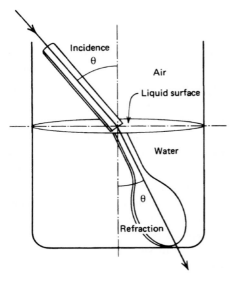

FIGURE 29.1   The refraction of light.

FIGURE 29.2 One model of a refractometer.

3. **Hinged sample prisms.** This is where you put your sample.

4. **Light source.** This provides light for your sample. It's on a movable arm, so you can swing it out of the way when you place your samples on the prisms.

5. **Light source swivel-arm pivot and lock.** This is a large, slotted nut that works itself loose as you move the light source up and down a few times. Always have a dime handy to help you tighten this locking nut when it gets loose.

6. **Scale adjust knob.** You use this knob to adjust the optics so that you see a split field in the eyepiece. The refractive index scale also moves when you turn this knob. The knob is often a dual control; use the outer knob for a coarse adjustment and the inner knob for a fine adjustment.

7. **Scale/sample field switch.** Press the "Press and hold to read scale" switch, and the numbered refractive index scale appears in the eyepiece. Release this switch, and you see your sample in the eyepiece. Some models don't have this type of switch. You have to change your angle of view (shift your head a bit) to see the field with the refractive index reading.

8. *Line cord on–off switch.* This turns the refractometer light source on and off.

9. *Recirculating water inlet and outlet.* These are often connected to temperature-controlled recirculating water baths. The prisms and your samples in the prisms can all be kept at the temperature of the water.

BEFORE USING THE ABBÉ REFRACTOMETER: A LITTLE PRACTICE

This is going to sound very strange, but before you use the refractometer to get the refractive index of your unknown, use the refractometer on a common known—**water**! See, there are so many things you can vary—lamp position, compensating prisms, scale adjustment—well, if you've never used this refractometer, you could misadjust, say, the lamp position and totally miss the change in the field and drive the scale adjustment past the point of no return. If you practice with water, you remove one of the variables.

1. Open the prisms, make sure they're clean, and put water on the lower prism. Then close the prisms.

2. Now, press and hold the scale/sample field switch, and, looking at the upper-most scale, carefully, purposefully set the instrument to read 1.3333, the index of refraction of water.

3. Release the scale/sample field switch, and now start at step 3 in the next section, "Using the Abbé Refractometer," to see the effects of moving the lamp position and the compensating prism adjustment drum.

USING THE ABBÉ REFRACTOMETER

1. Make sure the unit is plugged in. Then turn the on–off switch to ON. The light at the end of the movable arm should come on.

2. Open the hinged sample prisms. *Without touching the prisms at all,* place a few drops of your liquid on the lower prism. Then swing the upper prism back over the lower one and *gently* close the prisms. *Never touch the prisms with any hard object, or you'll scratch them.*

3. Raise the light on the end of the movable arm so that the light illuminates the upper prism. Get out your dime and, with the permission of your instructor, tighten the light-source swivel arm locknut, as it gets tired and lets the light drop.

4. Look in the eyepiece. Slowly, carefully, with *very little* force, turn the large scale and sample image adjust knob from one end of its rotation to the other. *Do not force!* (If your sample is supposedly the same as that of the last person to use the refractometer, you shouldn't have to adjust this much, if at all.)

| Too high | Too low | Just right |

— No color fringes

FIGURE 29.3 Your sample through the lens of the refractometer.

5. Look for a split optical field of light and dark (Fig. 29.3). This may not be very distinct. You may have to raise or lower the light source and scan the sample a few times.

6. If you see color fringing at the boundary between light and dark (usually red or blue), slowly turn the compensating prism adjust drum until the colors are at a minimum.

7. Use the image adjust knob to center the split optical field right on the cross-hairs.

8. Press and hold the scale/sample field switch. The refractive index scale should appear (Fig. 29.4). Read the uppermost scale, the refractive index, to four decimal places (five digits here). (If your model has two fields, and the refractive index is *always* visible, just read it.)

9. Read and record the temperature of the refractometer. You'll have to correct the reading for any difference between the temperature of your sample and the temperature given in your literature reference. If you don't have that value, correct the reading to 20°C. The correction is 0.0004 units per degree difference above the reference value.

10. Switch the unit to OFF. The swing-arm light should go off.

11. Clean both prisms with ethanol and a Kimwipe. Dispose of waste properly.

1.440

FIGURE 29.4 A refractive index of 1.4397.

REFRACTOMETRY HINTS

1. The refractive index changes with temperature. If your reading is not the same as that of a handbook, check the temperatures before you despair.

2. Volatile samples require quick action. Cyclohexene, for example, has been known to evaporate from the prisms of unthermostatted refractometers more quickly than you can obtain the index. It may take several tries as you readjust the light, turn the scale and sample image adjust, and so on.

3. Make sure the instrument is level. Often, organic liquids can seep out of the prism jaws before you are ready to make your measurement.

Gas chromatography (GC) can also be referred to as **vapor-phase chromatography** (VPC) and even **gas–liquid chromatography** (GLC). Usually, the technique, the instrument, and the recording of the data are called **GC:**

"Fire up the GC."   (the instrument)
"Analyze your sample by GC." (perform the technique)
"Get the data off the GC."   (analyze the chromatogram)

I've mentioned the similarity of all chromatography, and just because electronic instrumentation is used, there's no need to feel that something basically different is going on.

THE MOBILE PHASE: GAS

In **column chromatography,** the mobile (moving) phase is a liquid that carries your material through an adsorbent. I called this phase the **eluent,** remember? Here, a gas is used to push, or *carry,* your vaporized sample, and it's called the **mobile phase.**

The **carrier gas** is usually helium, although you can use nitrogen. You use a **microliter syringe** to inject your sample into this gas stream through an **injection port** and then *onto the column*. If your sample is a mixture, the *compounds separate on the column* and reach the **detector** at different times. As each component hits the detector, the detector generates an **electrical signal.** Usually, the signal goes through an **attenuator network,** and then out to a **computer** to record the signal. I know that's a fairly general description, and Figure 30.1 is a highly simplified diagram. But there are lots of different GCs, so being specific about their operation doesn't help here. You should see your instructor. But that doesn't mean we can't talk about some things.

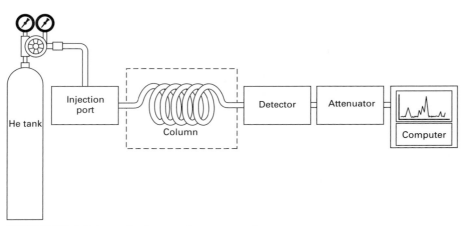

FIGURE 30.1 Schematic of a gas chromatography setup.

GC SAMPLE PREPARATION

Sample preparation for GC doesn't require much more work than handing in a sample to be graded. *Clean and dry,* right? Try to take care that the boiling point of the material is low enough that you can actually use this technique. The maximum permissible temperature depends on the type of column, and that should be given. In fact, for any single experiment that uses GC, the nature of the column, the temperature, and most of the electronic settings will be fixed.

GC SAMPLE INTRODUCTION

The sample enters the GC at the **injection port** (Fig. 30.2). You use a microliter syringe to pierce the rubber septum and inject the sample onto the column. Don't stab yourself or anyone else with the needle. Remember, this is *not* dart night at the pub. Don't throw the syringe at the septum. There is a way to do this.

1. To load the sample, put the needle into your liquid sample, and *slowly* pull the plunger to draw it up. If you move too fast and more air than sample gets in, you'll have to push the plunger back again and draw it up once more. Usually they give you a 10-μL syringe, and 1 or 2 μL of sample is enough. Take the loaded syringe out of the sample, and *carefully, cautiously* pull the plunger back so that there is no sample in the needle. You should see a bit of air at the very top but not very much. This way you don't run the risk of having your compound boil out of the needle as it enters the injector oven just before you actually inject your sample. That premature injection broadens the sample and reduces the resolution. In addition, the air acts as an internal standard. Since air travels through the column almost as fast as the carrier gas, the **air peak**

FIGURE 30.2 A GC injection port chromatography setup.

that you get can signal the start of the chromatogram, much like the notch at the start of a TLC plate. Ask your instructor.

2. Hold the syringe in *two hands* (Fig. 30.3). There is no reason to practice being an M.D. in the organic laboratory.

3. Bring the syringe to the level of the injection port, straight on. No angles. Then let the needle touch the septum at the center.

4. The real tricky part is holding the barrel and, without injecting, pushing the needle through the septum. This is easier to write about than it is to do the first time. Pierce the septum carefully.

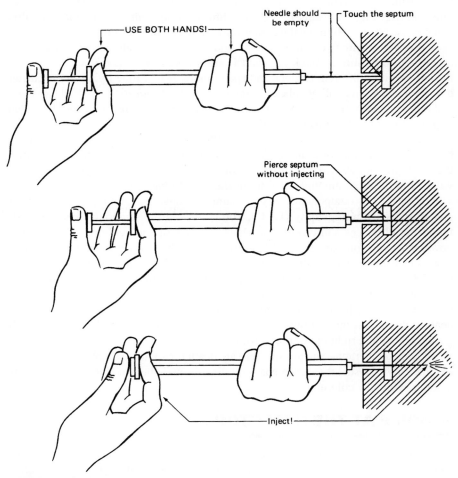

FIGURE 30.3 Three little steps to a great GC.

5. Now quickly and smoothly push on the plunger to inject the sample, and then pull the syringe needle out of the septum and injection port.

After a while, the septum gets full of holes and begins to leak. Usually, you can tell that you have a leaky septum when the trace on the computer wanders about aimlessly without any sample injected.

Speaking of computers, as soon as you do inject that sample, please tell your computer that you have done so. It's been waiting, wagging its tail like a happy puppy, for you to give it the signal to run.

SAMPLE IN THE COLUMN

Now that the sample is in the column, you might want to know what happens to this mixture. Did I say mixture? Sure. Just as with **thin-layer** and **column chromatography,** you can use GC to determine the composition or purity of your sample.

Let's start with two components, A and B again, and follow their path through an **adsorption column.** Well, if A and B are different, they are going to stick on the adsorbent to different degrees and spend more or less time flying in the carrier gas. Eventually, one will get ahead of the other. Aha! *Separation*—just like column and thin-layer chromatography. Only here the samples are vaporized, and it's called **vapor-phase chromatography (VPC).**

Some of the adsorbents are coated with a **liquid phase.** Most are very high-boiling liquids, and some look like waxes or solids at room temperature. Still, they're liquid phases. So, the different components of the mixture you've injected will spend different amounts of time in the liquid phase, and, again, you'll get a separation of the components in your sample. Thus, the technique is known as **gas–liquid chromatography (GLC).** You could use the same adsorbent and different liquid phases and change the characteristics of each column. Can you see how the sample components would partition themselves between the gas and liquid phases and separate according to, perhaps, molecular weight, polarity, size, and so on, making this technique also known as **liquid-partition chromatography?**

Since these liquid phases on the adsorbent are eventually liquids, you can boil them. And that's why there are temperature limits for columns. It is not advisable to heat a column past the recommended temperature, boiling the liquid phase right off the adsorbent and right out of the instrument.

High temperature and air (oxygen) are death for some liquid phases, since they oxidize. So make sure the carrier gas is running through them at all times, even a tiny amount, while the column is hot.

SAMPLE AT THE DETECTOR

There are several types of **detectors,** devices that can tell when a sample is passing by them. They detect the presence of a sample and convert it to an electrical signal that's turned into a **GC peak** (Fig. 30.4) on the chart recorder. The most common type is the **thermal conductivity detector.** Sometimes called hot-wire detectors,

FIGURE 30.4 A well-behaved GC trace of a mixture of three compounds.

	A	B	C
Height (mm)	8	68	145
Width at half-height (mm)	4	4	4
Area (mm²)	32	212	580
Relative area	1	8.5	18.1
Distance from injection (in.)	3	3⁵/₁₆	3¹¹/₁₆
Retention times t_R at 2 min/in. (min)	6	6.63	7.38

these devices are very similar to the filaments you find in lightbulbs, and they require some care. Don't ever turn on the **filament current** unless the carrier gas is flowing. A little air (oxygen), a little heat, a little current, and you get a lot of trouble replacing the burned-out detector.

Usually, there are at least two thermal conductivity detectors in the instrument, in a "bridge circuit." Both detectors are set in the gas stream, but only one sees the samples. The electric current running through them heats them up, and they lose heat to the carrier gas at the same rate.

As long as no sample, only carrier gas, goes over *both* detectors, the bridge circuit is **balanced.** There is no signal to the computer, and the trace or value displayed does not move from zero.

Now a sample in the carrier gas goes by one detector. This sample has a thermal conductivity different from that of pure carrier gas. So the **sample detector** loses heat at a different rate from the **reference detector.** (Remember, the **reference** is the detector that NEVER sees samples—only carrier gas.) The detectors are in different surroundings. They are not really equal anymore. So the bridge circuit becomes unbalanced, and a signal goes to the computer, giving a GC peak.

Try to remember the pairing of **sample** with **reference;** it's the *difference* between the two that most electronic instrumentation responds to. You will see this again and again.

ELECTRONIC INTERLUDE

The electrical signal makes two other stops on its way to the computer.

1. *The coarse attenuator.* This control makes the signal weaker (attenuates it). Usually, there's a scale marked in **powers of two:** 2, 4, 8, 16, 32, 64, . . . , so each position is half as sensitive as the last one. There is one setting, either an ∞ or an **S** (for **shorted**), which means that the attenuator has *shorted out the terminals connected to the computer.* Now the **computer software zero** can be set properly.

2. *The GC zero control.* This control helps set the zero position on the computer, but it is *not to be confused with the zero control in the computer software.*

Here's how to set up the electronics properly for a GC and a data-collection program.

1. The computer should be booted up, and the GC data collection software started. The **GC** should be allowed to warm up and stabilize for at least 10–15 minutes. Some systems take more time; ask your instructor.

2. Set the **coarse attenuator** to the highest attenuation, usually an ∞ or an S. With the GC output shorted, no signal reaches the computer.

3. Now, set the computer readout—either a trace or a voltage value—to zero, using *only the computer software zero control.* Once you do that, *leave it alone.* This way, you're not introducing a zero-offset error—a reading either added to or subtracted from the signal from the GC by the computer.

4. Start turning the *coarse attenuator* control to more sensitive settings (lower numbers), and *watch the trace on the computer.*

5. If the computer reading moves off zero, *use only the GC zero control* to bring the trace back to the zero point on the computer screen. *Do not use the computer software to get a reading of zero. Use only the GC zero control.*

6. As the *coarse attenuator* reaches more sensitive settings (lower numbers), it becomes more difficult to adjust the *trace to zero* using *only the GC zero control.* Do the best you can at the *lowest attenuation* (highest sensitivity) you can hold a zero steady at.

7. Now, you *don't* normally run samples on the GC at attenuations of 1 or 2. These settings are *very sensitive,* and there may be lots of electrical noise—the trace jumps about. The point is, *if the GC zero is OK at an attenuation of 1,* then when you run at attenuations of 8, 16, 32, and so on, *the baseline will not jump if you change attenuation in the middle of the run.*

Now that the attenuator is set to give peaks of the proper height, you're ready to go. Just be aware that there may be a **polarity switch** that can make your peaks shift direction.

SAMPLE ON THE COMPUTER

Basic computer analysis of a GC trace can be pretty simple:

1. *Mark your starting point.* This is the same as that mark you put on TLC plates to show where you spotted your sample. If you have an air peak, set your starting point there. If you don't have an air peak, the starting point would have to be when you first had the computer start collecting data. Can you set the starting point anywhere else? Some software packages let you set the start anywhere during a run. For the basic analysis of getting the retention time, I'm not sure how useful that would be.

2. *Mark the peaks.* Use the software to give the retention times. Usually you select the top of a peak, give a mouse click, and behold—the retention time. If you used an air peak, you can get the software to say that the air peak is at time $= 0.0$, and go from there.

3. *Get the areas.* Sometimes, right after you have identified the peaks you want by selecting them, the software is ready to print out the relative peak areas right away. In a GC trace with a good, stable baseline, this is probably all you need. If you have baseline drift, or ill-resolved (partially overlapping) peaks, the simple, automatically calculated values might not do the job.

4. *Massage the areas.* Again, you might not have to do this with a well-behaved GC trace. Otherwise, you'll have to use special baseline-correction functions, or even peak-separation functions, in the software to get more accurate results. For baseline correction, you might have to draw or click a baseline under a particular peak yourself before letting the computer calculate the area under the peak.

With peak areas you can calculate how much of each compound is in your sample.

PARAMETERS, PARAMETERS

To get the best GC trace from a given column, there are lots of things you can do, simply because you have so many controls. Usually you'll be told the correct conditions, or they'll be preset on the GC.

Gas Flow Rate

The faster the carrier gas flows, the faster the compounds are pushed through the column. Because they spend *less time in the stationary phase,* they don't separate as well, and the *GC peaks come out very sharp but not well separated.* If you *slow the carrier gas down too much,* the compounds spend so much time in the stationary phase that *the peaks broaden and the overlap gets very bad.* The optimum is, as always, the best separation you need, in the shortest amount of time. Sometimes you're on your own. Most of the time, someone else has already worked it out for you.

Temperature

Whether or not you realize it, the GC column has its own heater—the **column oven.** If you turn the temperature up, the compounds hotfoot it through the column very quickly. Because they spend *less time in the stationary phase,* they don't separate as well, and the *GC peaks come out very sharp but not well separated.* If you *turn the temperature down some,* the compounds spend so much time in the stationary phase that *the peaks broaden and overlap gets very bad.* The optimum is, as always, the best separation you need in the shortest amount of time. There are two absolute limits, however.

1. *Too high a temperature, and you destroy the column.* The adsorbent may decompose, or the liquid phase may boil out onto the detector. *Never exceed the recommended maximum temperature for the column material.* Don't even come within 20°C of it.

2. *Way too low a temperature, and the material condenses on the column.* The temperature has to be above the **dew point** of the least volatile material. Not the boiling point. Water doesn't always condense on the grass—*become dew*—every day that's just below 100°C (that's 212°F, the boiling point of water). Fortunately, you don't have to know the dew points for your compounds. However, you *do* have to know that the temperature doesn't *have to be above the boiling point* of your compounds.

Incidentally, the **injector** may have a separate **injector oven,** and the **detector** may have a separate **detector oven.** Set them both 10–20°C higher than the column temperature. You can even set these *above the boiling points* of your compounds, since you *do not want them to condense in the injection port or the detector, ever.* For those with *only one temperature control*, sorry. The injection port, column, and detector are all in the same place, all in the same oven, and all at the same temperature. The *maximum temperature,* then, is *limited by the decomposition temperature of the column.* Fortunately, because of that dew point phenomenon, you really don't have to work at the boiling points of the compounds.

EXERCISES

1. How do the temperature and the flow rate of the carrier gas affect the retention time of two separable compounds? Does the amount of separation also change? If so, how?

2. If air can degrade a column and cause TC detector filaments to burn out, why do we use an air peak?

HPLC. Is it **high-performance liquid chromatography** or **high-pressure liquid chromatography,** or something else? It's probably easier to consider it a delicate blend of **wet-column chromatography** and **gas chromatography** (see Chapters 28 and 30, respectively).

Rather than letting gravity **pull** the solvent through the powdered adsorbent, the liquid is **pumped** through under pressure. Initially, high pressures [1000–5000 psig (**pounds per square inch gauge**), i.e., not absolute] were used to push a liquid through a tightly packed solid. But the technique works well at lower pressures (250 psig); hence the name high-performance liquid chromatography.

From there on, the setup (Fig. 31.1) closely resembles gas chromatography. Although a **moving liquid phase** replaces the helium stream, compounds are put onto a column through an **injection port.** They *separate inside a chromatographic column* in the same way as in GC by spending more or less time in a *moving liquid,* and the separated compounds pass *through a detector*. There, as the amounts of each compound go by the **detector,** they are turned into electrical signals and *displayed on a computer screen as HPLC peaks* that look just like GC peaks. You should also get the feeling that the analysis of these HPLC traces is done similarly to GC traces, because it is.

Again, there's a lot of variety among HPLC systems, so what I say won't necessarily apply to your own system in every respect. But it should help.

THE MOBILE PHASE: LIQUID

If you use only *one liquid,* either **neat** or as a mixture, the entire chromatogram is said to be **isochratic**. Units that can deliver *varying solvent compositions over time* are called **gradient elution systems.**

For an **isochratic** system, you usually use a single solution, or a neat liquid, and put it into the **solvent reservoir,** which is generally a glass bottle that has a stopcock at the bottom to let the solvent out (Fig. 31.2). The solvent travels out of the bottom of the reservoir and usually through a **solvent filter** that traps any fairly large, insoluble impurities in the solvent.

If you're making up the eluent yourself, it is important to follow the directions *scrupulously*. Think about it. If you wet the entire system with the wrong eluent, you can wait a very long time for the *correct* eluent to reestablish the correct conditions.

A Bubble Trap

Air bubbles are the nasties in HPLC work. They cause the same type of troubles as they do in **wet-column chromatography,** and you just don't want them. So there's usually a **bubble trap** (Fig. 31.3) before the eluent reaches the pump. This device

FIGURE 31.1 Block diagram of an HPLC setup.

is really quite simple. Bubbles in the eluent stream rise up the center pipe and are trapped there. To get rid of the bubbles, you open the cap at the top. Then, solvent rises in the tube and pushes the bubbles out. You have to be *extremely* careful about bubbles if you're the one to start the setup or if the solvent tank has run out. Normally, *one bubble purge per day* is enough.

FIGURE 31.2 Aspirator bottle used to deliver eluent.

The Pump

The most common pumping system is the **reciprocating pump.** Milton Roy makes a pretty good model. The pump has a **reciprocating ruby rod** that moves back and forth. On the backstroke, the pump loads up on a little bit of solvent; then, on the forward stroke, it squirts it out under pressure. If you want to increase the amount of liquid going through the system, you can dial the length of the stroke from zero to a preset maximum by using the micrometer at the front of the pump (see Figure 31.4). Use a *fully clockwise* setting, and the stroke length is zero—*no solvent flow.* A *fully counterclockwise* setting gives the maximum stroke length—*maximum solvent flow.* If you have a chance to work with this type of pump, *always turn the micrometer*

FIGURE 31.3 Common bubble trap.

FIGURE 31.4   Pump, pulse dampener, and pressure gauge unit.

fully clockwise to give a zero stroke length before you start the pump. If there *is* a stroke length set *before* you turn the pump on, the first smack can damage the reciprocating ruby rod. And it is *not* cheap.

The Pulse Dampener

Because the rod **reciprocates** (i.e., goes back and forth), you'd expect huge swings in pressure, and pulses of pressure do occur. That's why they make **pulse dampeners.** A coiled tube is hooked to the pump on the side opposite the column (Fig. 31.4). It is filled with the eluent that's going through the system. On the forward stroke, solvent is compressed into this tube, and simultaneously a shot of solvent is pushed onto the column. On the backstroke, while the pump chamber fills up again, the eluent we just pressed into the pulse dampener *squirts out into the column.* Valves in the pump take care of directing the flow. As the eluent in the pulse dampener tubing takes up the slack, the huge variations in pressure, from essentially zero to perhaps 1000 psig, are evened out. They don't disappear, going about 100 psig either way, but these **dampened pulses** are now too small to be picked up on the detector. They don't show up on the chart recorder either.

HPLC SAMPLE PREPARATION

Samples for HPLC must be *liquids* or *solutions*. It would be nice if the solvent in which you've dissolved your solid sample were the same as the eluent.

It is *absolutely crucial* that you preclean your sample. *Any* decomposed or insoluble material will stick to the top of the column and can continually poison further runs. There are a few ways to keep your column clean.

1. *The Swinney adapter* (Fig. 31.5). This handy unit locks onto a syringe *already* filled with your sample. Then you push the sample *slowly* through a Millipore filter to trap insoluble particles. This does *not,* however, get rid of **soluble tars** that can ruin the column. (Oh: Don't confuse the Millipore filters with the papers that separate them. It's embarrassing.)

FIGURE 31.5   The Swinney adapter and syringe parts. (Note Luer-Lok.)

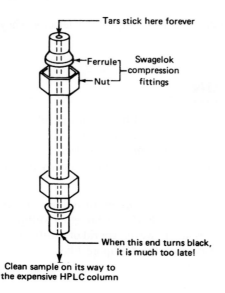

Tars stick here forever

Ferrule

Swagelok
—compression
fittings

Nut

When this end turns black,
it is much too late!

Clean sample on its way to
the expensive HPLC column

FIGURE 31.6 The precolumn filter.

2. *The precolumn filter* or *guard column* (Fig. 31.6). Add a tiny column filled
   with exactly the same material as the main column, and *let this small column
   get contaminated.* After a while, it used to be that you removed it, cleaned out
   the gunk and adsorbent, and refilled it with fresh column packing. Today it just
   gets thrown out. Either way, you don't really know when the garbage is going to
   poison the entire precolumn filter and then start ruining the analyzing column.
   The only way to find out whether you have to clean the precolumn is to take it
   out of the instrument. You really want to clean it out or throw it out long before
   the contaminants start to show up at the precolumn, or guard column, exit.

HPLC SAMPLE INTRODUCTION

This uses the equivalent of the **injection port** for the GC technique. With GC, you
could inject through a rubber septum directly onto the column. With HPLC, it's very
difficult to inject against a liquid stream moving at possibly 1000 psig. That's why
injection port valves were invented for HPLC: You put your sample into an **injection loop** on the valve that is *not in the liquid stream,* then *turn the valve,* and *voilà,
your sample is in the stream,* headed for the column.

The injection port valve (Fig. 31.7) has two positions:

1. *Normal solvent flow.* In this position, the eluent enters the valve, goes around,
   and comes on out into the column without any bother. *You put the sample in
   the injection loop in this position.*

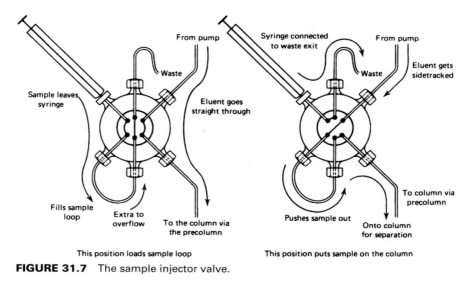

From pump

Syringe connected
to waste exit

From pump

Waste

Eluent gets
sidetracked

Sample leaves
syringe

Eluent goes
straight through

Waste

Fills sample
loop

Extra to
overflow

To the column via
the precolumn

Pushes sample out

Onto column
for separation

To column via
precolumn

This position loads sample loop

This position puts sample on the column

FIGURE 31.7   The sample injector valve.

2. *Sample introduction.* Flipped this way, the valve allows the eluent to be *pumped through the injection loop,* and any sample there is carried along and into the column. *You put the sample on the column in this position.*

SAMPLE IN THE COLUMN

Once the sample is in the column, there's not much difference between what happens here and what happens in paper, thin-layer, vapor-phase (gas), or wet-column chromatography. *The components in the mixture will remain in the stationary phase or move in the mobile phase for different times and end up at different places when you stop the experiment.*

So what's the advantage of HPLC? You can separate and detect microgram quantities of solid samples much as in GC, but for GC your solid samples have to vaporize without decomposing. Fat chance of that. You can analyze a solid sample without decomposing it with HPLC.

HPLC also uses something called a **bonded reversed-phase column,** in which the stationary phase is a nonpolar hydrocarbon that is chemically bonded to a solid support. You can use these with aqueous eluents, usually alcohol–water mixtures. So you have a *polar eluent and a nonpolar stationary phase,* something that does not usually occur for ordinary wet-column chromatography. One advantage is that you don't need to use anhydrous eluents with reversed-phase columns (very small amounts of water can change the character of normal-phase columns).

SAMPLE AT THE DETECTOR

Many HPLC detectors can turn the presence of your compound into an electrical signal to be acquired by a computer software program. Previously, **refractive index** detectors were common. *Clean eluent*, used as a reference, went through one side of the detector, and the *eluent with the samples* went through the other side. *A difference in the refractive index between the sample and reference generated an electrical signal that was sent to a chart recorder.* If you've read Chapter 30 on **gas chromatography** and looked ahead to Chapter 32 on **infrared spectroscopy,** you shouldn't be surprised to find both a **sample** and a **reference.** I did tell you the reference/sample pair is common in instrumentation.

More recently, **UV detectors** (Fig. 31.8) have become more popular. UV radiation is beyond the purple end of the rainbow, the energy from which great tans are made. So, if you set up a **small mercury-vapor lamp** with a power supply to light it up, you'll have a **source of UV light.** It's usually filtered to let through only the wavelengths of 180 and/or 254 nm. (And *where* have you met *that* number before? TLC plates, maybe?) This UV light then passes through a **flow cell** that has the *eluent and your separated sample flowing through it* and uses air as a reference. *When your samples come through the cell, they absorb the UV, and an electrical signal is generated.* Yes, the signal goes to a computer and shows up as HPLC peaks.

FIGURE 31.8 Cutaway view of an HPLC UV detector.

SAMPLE ON THE COMPUTER

Go back and read about HPLC peak interpretation in the sections on GC detector electronic setup ("Electronic Interlude" in Chapter 30) and peak interpretation ("Sample on the Computer" in Chapter 30). The analysis is *exactly the same—* retention times, peak areas, baselines . . . all that.

PARAMETERS, PARAMETERS

To get the best HPLC trace from a given column, there are lots of things you can do; most of them are the same as those for GC (see "Parameters, Parameters" in Chapter 30).

Eluent Flow Rate

The faster the eluent flows, the faster the compounds are pushed through the column. Because they spend *less time in the stationary phase,* they don't separate as well, and the *HPLC peaks come out very sharp but not well separated.* If you *slow the eluent down too much,* the compounds spend so much time in the stationary phase that the *peaks broaden and overlap gets very bad.* The optimum is, as always, the best separation you need in the shortest amount of time. One big difference in HPLC is the need to worry about **back-pressure.** If you try for very high flow rates, the HPLC column packing tends to collapse under the pressure of the liquid. This, then, is the cause of the back-pressure: *resistance of the column packing.* If the pressures get *too high,* you may burst the tubing in the system, damaging the pump . . . all sorts of fun things.

Temperature

Not many HPLC setups have ovens for temperatures like those for GC because *eluents tend to boil at temperatures much lower than the compounds on the column,* which are usually solids anyway. Eluent bubbling problems are bad enough without actually boiling the solvent in the column. This is not to say that HPLC results are independent of temperature. They're not. But if a column oven for HPLC is present, its purpose more likely is to keep stray drafts and sudden chills away than to have a hot time.

Eluent Composition

You can vary the composition of the eluent (mobile phase) a lot more in HPLC than in GC, so there's not really much correspondence. Substitute nitrogen for helium in GC, and usually the sensitivity decreases, but the retention times stay the same. Changing the mobile phases—the gases—in GC doesn't have a very big effect on the separation or retention time.

There are much better parallels to HPLC: **TLC** or **column chromatography.** Vary the eluents in these techniques, and you get widely different results. With a **normal-phase, silica-based column,** you can get results similar to those from silica gel TLC plates.

EXERCISES

1. Compare and contrast HPLC with GC.

2. Gradient elution, where the solvent changes as the HPLC is run (much like temperature programming in GC), is useful in what kinds of cases? Someone programmed the HPLC to start with the most polar solvent and then switch over to a less polar one, the reverse of the usual case. Is this an error, or are there applications where this might be useful?

INFRARED SPECTROSCOPY

(and a bit of UV-VIS, too)

In **infrared (IR) absorption spectroscopy,** infrared radiation of wavelengths from 2.5 to 15 micrometers (μm; microns) gets passed through (usually) or reflected from your product, and, as the instrument scans your sample, *specific functional groups absorb at specific energies.* Because this spectrum of energies is laid out on a piece of paper, these specific energies become specific places on a chart.

MOLECULES AS BALLS ON SPRINGS

You can model a diatomic molecule as two balls of mass m connected by a spring with a force constant (stiffness) k. This is called the **classical harmonic oscillator** that we know and love from physics. At equilibrium distance (d), the displacement from this position (x) is zero, and the spring has no potential energy. Stretch the spring to $+x$, or compress the spring to $-x$, and the energy increases along the curve. That curve is a parabola called the **harmonic potential** (Fig. 32.1).

If you get this system going from $+x$ through zero to $-x$ and back, again and again, you've set up a vibration. And the frequency (v) of that vibration is given by the formula

$$v = \frac{1}{2\pi}\sqrt{\frac{k}{m}} \tag{1}$$

Strictly speaking, for a system of two masses connected by a spring, m here is replaced by the reduced mass μ (mu):

$$\mu = \frac{m_1 \cdot m_2}{m_1 + m_2} \tag{2}$$

Nonetheless, equation (1) shows that if you have either a stronger spring (larger force constant, k) or a smaller mass (m), then the frequency of oscillation is going to go up. A weaker spring or a larger mass, and the frequency of the vibration will go down.

FIGURE 32.1 Harmonic potential of a simple ball-and-spring oscillator.

Total energy
$E = \frac{1}{2} kx^2$ max *or* min
Any *x*; any *E*

x_{min} x_{max}

FIGURE 32.2   An unrestricted position for *x* gives any *E*.

This ball-and-spring model will vibrate with a total energy (E) equal to the potential energy at the stretched or compressed position (Fig. 32.2):

$$E = \frac{1}{2} kx^2_{\text{max } or \text{ min}} \tag{3}$$

While balls on springs can vibrate at any energy (and any frequency), molecules, even simple diatomic ones, don't have this luxury.

AH, QUANTUM MECHANICS

Quantum mechanics predicts that the molecule may vibrate only at energy levels that fit the formula

$$E_n = \left(n + \frac{1}{2} \right) hv \qquad n = 0, 1, 2, 3 \tag{4}$$

where v is the frequency of vibration and h is Planck's constant (6.6262×10^{-34} joule-second). Because of the limitation to discrete energy levels, the energy is said to be **quantized.** The molecule can only absorb (or emit) light of energy equal to the spacing between two levels, and you can only go from one level to the next (the **selection rule** here is $\Delta n = \pm 1$) (Fig. 32.3).

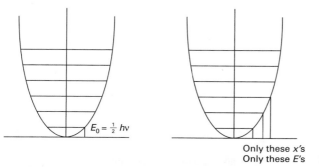

$E_0 = \frac{1}{2} hv$

Only these x's
Only these E's

FIGURE 32.3   Quantization of energy levels sets restrictions.

For the vibrational levels for ordinary molecules, the necessary and sufficient energy to effect a transition from one level to the next is contained in light energy whose frequencies are in the range of the infrared, approximately 4000 to 650 wavenumbers (wavelengths of about 2.5 to 16 μm). More on this later.

THE DISSONANT OSCILLATOR

Unfortunately, the absorptions of real diatomic molecules are much more complicated because they don't behave as harmonic oscillators. Pull the atoms apart too much, and you break the spring—er, bond. Push them together, and they start to repel each other; the energy of the system goes way up very rapidly. This behavior can be modeled with an **anharmonic potential**, and the vibrating system is usually called an **anharmonic oscillator.** The energy levels aren't equally spaced anymore; you can get transitions of more than one level now, and you can get more than one absorption in the infrared energy spectrum. The transition from $n = 0$ to $n = 1$ is called the **fundamental absorption**, just as before. Now that you can also get transitions from $n = 0$ to $n = 2$ (a different **selection rule**; $\Delta n = \pm 2$), you can get **overtone absorptions** that are almost—not quite—twice the frequency of the fundamental absorption. Why not exactly twice? For the anharmonic oscillator, the energy levels are no longer evenly spaced, so the energies for the transition from $n = 0$ to $n = 1$ won't be the same as those for the transition from $n = 1$ to $n = 2$ (Fig. 32.4).

BUT WAIT! THERE'S MORE

Even with all this, if there's no change in the **dipole moment**, the molecule won't absorb light. If the two masses on the ends of your spring (bond) are hydrogen atoms, and H_2 always has a net dipole of zero, there are no absorptions of light in the infrared spectrum. Hydrogen chloride (HCl) does have a dipole change as it stretches, and, as long as the frequency of the light is correct, will absorb infrared energy. The intensity of that absorption is related directly to the magnitude of that dipole change.

FIGURE 32.4 An anharmonic oscillator allows overtones as well as fundamental vibration.

MORE COMPLICATED MOLECULES

The water molecule is apparently not that much more complicated than a diatomic molecule, having only one more atom, but an analysis of the infrared absorption spectrum requires some care. With three atoms and two bonds, a water molecule can appear to be moving around more than those Spandex-clad folk on an exercise video.

The vibrations can be broken down into three simpler motions called **normal modes of vibration.** For water, there are the bend, the symmetric stretch, and the asymmetric stretch (often called the antisymmetric stretch). Every nonlinear molecule has $3N - 6$ normal modes of vibration, where N is the number of atoms in the molecule. Since each normal mode in the water molecule changes the dipole moment as it vibrates, each will absorb infrared light. Three peaks should appear in the infrared spectrum of water, with the two stretching vibrations overlapped because they are similar movements (Fig. 32.5).

If three atoms are part of a larger molecule—say, the —CH$_2$— groups in hexane—other bending motions can occur: in-plane scissoring and rocking, out-of-plane twisting and wagging, deep knee bends and squat thrusts—well, maybe not those last two (Fig. 32.6). Nonetheless, there will be a lot of different infrared absorptions for all but the simplest of molecules. Throw in overtone absorptions, and the direct assignment of these peaks to the particular motion of all the particular atomic groupings, and the task can seem formidable.

CORRELATION TABLES TO THE RESCUE

Ascribing meanings to all of the IR peaks in a typical IR spectrum can be very difficult. Just take a look at the spectra of *tert*-butyl alcohol or cyclohexanol (Fig. 32.7). That's why frequency correlation diagrams of IR tables exist. They identify regions of the IR spectrum where peaks for various functional groups show up. They can be graphical (Fig. 32.8) or tabular (Fig. 32.9), and you should try to get used to both styles when interpreting your infrared spectrum.

TROUGHS AND RECIPROCAL CENTIMETERS

Although the technique is called infrared absorption spectroscopy, you'll notice that the abscissa (y-axis) is specified in **percent transmission** (as in Figure 32.7) or **%T.** Although you can convert from one scale to the other,

$$Absorbance = 2 - \log_{10}\%\text{T},$$

the absorbance scale, going from 0 to ∞, can cause problems, especially for high values of absorption. Thus the vastly more well-behaved %T scale is used, and thus the "troughs." We still call them peaks, though. (As an exercise, get out your calculator or spreadsheet and solve for the absorbance with %T being zero, or a complete block of radiation.)

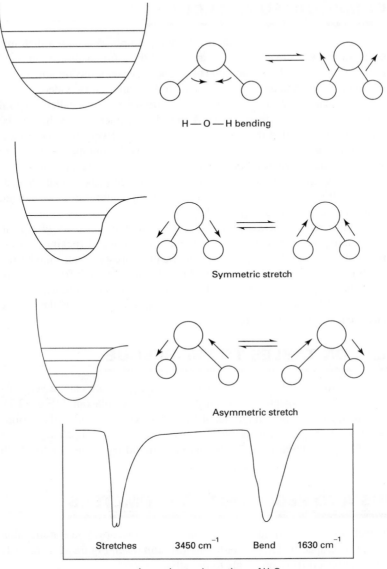

H — O — H bending

Symmetric stretch

Asymmetric stretch

Stretches   3450 cm^{-1} Bend   1630 cm^{-1}

Approximate absorptions of H_2O

FIGURE 32.5 Stretches and a bend give rise to the infrared absorption spectrum of water.

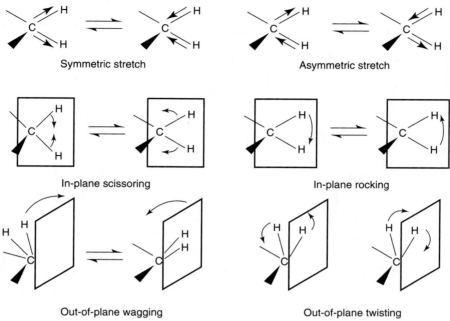

FIGURE 32.6 Some —CH₂— groups going through their paces.

Although wavelength (λ) in **micrometers (μm; microns)** can be used as the ordinate (*x*-axis) units to denote the position of peaks on the infrared spectrum, almost all assignments are given in something called **reciprocal centimeters,** or **wavenumbers (cm^{-1}).** Again, it's a simple calculation to convert from one into the other:

$$cm^{-1} = \frac{1}{\mu m} \times 10,000 \qquad \text{and} \qquad \mu m = \frac{1}{cm^{-1}} \times 10,000$$

In fact, spectral printouts often have both scales. If you look a little more carefully, you'll see that one scale—usually the wavenumber scale—is linear, and the other scale is not. It turns out that absorbances for most **functional groups** are generally in the first half of the spectrum (4000 to 1400 cm^{-1} or 2.5 to 7.2 μm), while the **fingerprint region** ranges from 1400 to 990 cm^{-1} or 7.2 to 11.1 μm. If you plot your spectrum **linear to wavenumbers,** then the first part, *the functional group region, gets expanded,* and the fingerprint region gets compressed. Plot your spectrum **linear to wavelength,** and the front part, *the functional group region, gets compressed* while the fingerprint region gets expanded. With the advent of computer data acquisition of spectral data, it's easy to shift from one to the other, so you can see which kind of plot will reveal the most pertinent information about your compound. Since the purpose of taking the infrared absorption spectrum of your compound is to identify the functional groups present, plots linear to wavenumbers will probably be more useful.

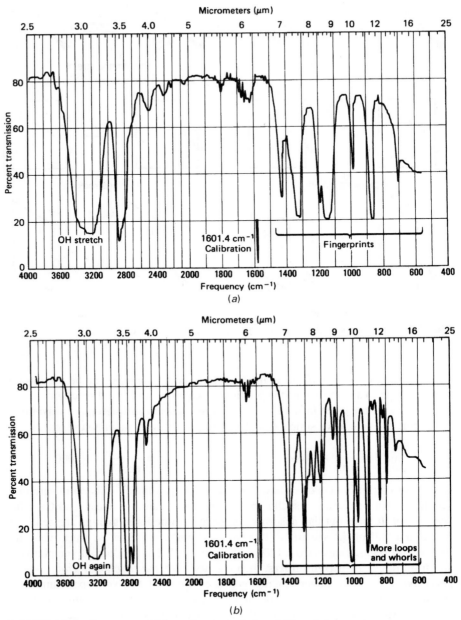

FIGURE 32.7   IR spectra of (a) *tert*-butyl alcohol and (b) cyclohexanol.

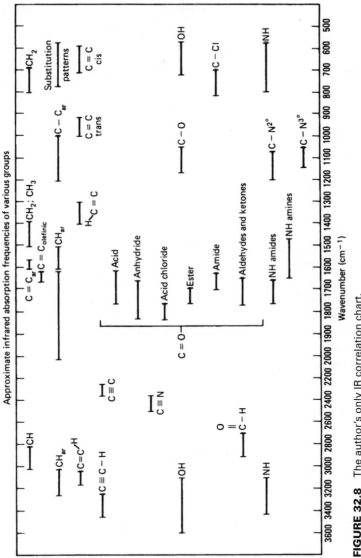

FIGURE 32.8 The author's only IR correlation chart.

Characteristic Infrared Absorptions of Functional Groups

Group	Frequency Range (cm^{-1})	Intensity[a]
A. Alkyl		
C—H (stretching)	2853–2962	(m—s)
Isopropyl, —CH(CH$_3$)$_2$	1380–1385 and 1365–1370	(s)
tert-Butyl, —C(CH$_3$)$_3$	1385–1395 and ≈1365	(s)
B. Alkenyl		
C—H (stretching)	3010–3095	(m)
C=C (stretching)	1620–1680	(v)
R—CH=CH$_2$	985–1000 and 905–920	(s)
R$_2$C=CH$_2$	880–900	(s)
cis-RCH=CHR	675–730	(s)
trans-RCH=CHR	960–975	(s)
C. Alkynyl		
≡C—H (stretching)	≈3300	(s)
C≡C (stretching)	2100–2260	(v)
D. Aromatic		
Ar—H (stretching)	≈3030	(v)
Aromatic substitution type (C—H out-of-plane bendings)		
Monosubstituted	690–710 and 730–770	(very s)
ortho-Disubstituted	735–770	(s)
meta-Disubstituted	680–725 and 750–810	(very s)
para-Disubstituted	800–840	(s)
E. Alcohols, Phenols, Carboxylic Acids		
OH (alcohols, phenols, dilute solution)	3590–3650	(sharp, v)
OH (alcohols, phenols, hydrogen bonded)	3200–3550	(broad, s)
OH (carboxylic acids, hydrogen bonded)	2500–3000	(broad, v)
F. Aldehydes, Ketones, Esters and Carboxylic Acids		
C=O stretch (generic)	1630–1780	(s)
aldehydes	1690–1740	(s)
ketones	1680–1750	(s)
esters	1735–1750	(s)
carboxylic acids	1710–1780	(s)
amides	1630–1690	(s)
G. Amines		
N—H	3300–3500	(m)
H. Nitriles		
C≡N	2220–2260	(m)

[a]Abbreviations: s = strong m = medium w = weak v = variable ≈ = approximately

FIGURE 32.9 A tabular IR correlation table. From T. W. G. Solomons and C. B. Fryhle, *Organic Chemistry*, 8th ed., Wiley, New York, 2004.

SOME FUNCTIONAL GROUP ANALYSIS

Look at Figure 32.7. Here's a fine example of a pair of alcohols if ever there was one. See the peak (some might call it a trough) at about 3400 cm^{-1} (2.9 μm)? That's due to the **OH group,** specifically the stretch in the O—H bond, the **OH stretch.**

Now consider a couple of ketones, 2-butanone and cyclohexanone (Fig. 32.10). There's no OH peak at about 3400 cm^{-1} (2.9 μm), is there? Should there be? *Of course not.* Is there an OH in 2-butanone? *Of course not.* But there is a C=O, and where's that? The peak is at about 1700 cm^{-1} (5.9 μm). It's *not there* for the alcohols, and it *is there* for the ketones. Right. You've just correlated or *interpreted* four IRs.

Because the first two (Fig. 32.7) have the *characteristic OH stretch of alcohols,* they might just be alcohols. The other two (Fig. 32.10) might be ketones because of the *characteristic* C=O *stretch* at 1700 cm^{-1} (5.9 μm) in each.

Take another look at the cyclohexanol and cyclohexanone spectra (Fig. 32.7b and Fig. 32.10b). Both show very different functional groups. Now look at the similarities, the simplicity, including the fingerprint region. Both are six-membered rings and have a high degree of symmetry. You should be able to see the similarities due to the similar structural features.

Two more things. First, watch your spelling and pronunciation; it's not "infared," OK? Second, most people I know use IR (pronounced "eye-are," not "ear") to refer to the technique, the instrument, and the chart recording of the spectrum:

"That's a nice new IR you have there."   (the instrument)

"Take an IR of your sample."   (perform the technique)

"Let's look at your IR and see what kind of   (interpret the resulting
compound you have."   spectrogram)

A SYSTEMATIC INTERPRETATION

To interpret an infrared spectrum, you'll need to see if the characteristic peaks for functional groups are present or absent. Finding an O—H stretch in the infrared spectrum of an alcohol you've been trying to prepare is wonderful. Finding that O—H stretch in an alkene you're trying to make from that alcohol is not so good, to put it mildly. One of the systematic ways of tackling the interpretation of an infrared spectrum is to break up the spectrum into smaller regions in a left-to-right fashion, and see what is or is not there.

> *Region: 4000–3000 cm^{-1}.* Really, the only peaks in this region are the O—H and N—H stretches. You could have an —OH, or a primary or secondary amine. The O—H stretch should be both strong and broad. The N—H stretch is usually broad, but probably ranges in intensity from medium to pitiful. And don't think "no amine" if there's nothing here; tertiary amines have no N—H at all.

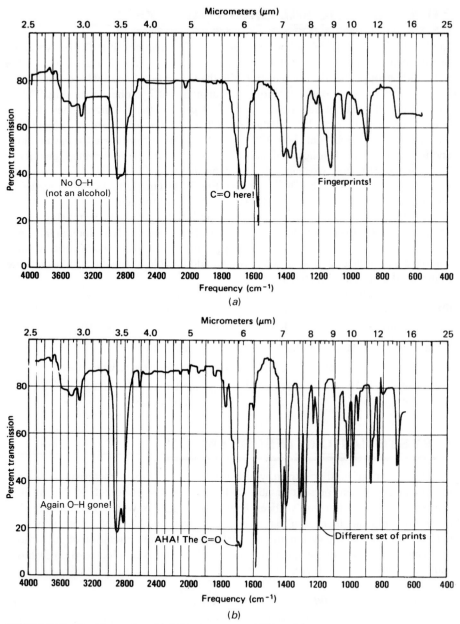

FIGURE 32.10 IR spectra of (a) 2-butanone and (b) cyclohexone.

Region: 3300–3000 cm^{-1}. The *sp* and *sp^2* C—H stretches are here. If you don't have them, it doesn't mean you don't have a C=C or a C≡C, just ones without hydrogens on them. You'll have to check for the C=C or C≡C stretches themselves; they can be weak and hard to find. The aromatic C—H stretch is found here as well. Actually, the entire spectrum of aromatic compounds tends to look "spiky" when compared to those of aliphatic compounds.

Region: 3000–2700 cm^{-1}. The *sp^3* C—H stretches are here. They're generally fairly sharp, well-defined, yet have overlapping peaks. If these C—H stretches appear to be on top of, or seem to disappear in, a very broad "swelling," check for a C=O peak farther down the line. That's the extremely hydrogen-bonded —OH of a carboxylic acid.

Region: 2250–2100 cm^{-1}. The C≡C and C≡N stretches are here. These tend to be medium-to-weak peaks, with C≡N being the strongest, terminal C≡C—H next, and finally internal C≡C.

Region: 1850–1600 cm^{-1}. This is where C=O stretches are, and they will be the strongest of absorptions. You'll have to do a little more sifting:
- A stretch around 1830–1800 cm^{-1}? You might have an acid chloride or, less likely, an acid bromide. If there is a second strong peak around 1775–1740 cm^{-1}, you have an anhydride.
- A stretch around 1700–1735 cm^{-1} could be the C=O of an aldehyde, a ketone, or a carboxylic acid. Got evidence of that —OH curve underlying the C—H stretches? Carboxylic acid. Without that, well, aldehydes should have evidence of a C—H stretch on their C=O at 2850–2700 cm^{-1}, a set of two small peaks that might be obscured by the C—H stretches in the region.
- A stretch around 1750–1730 cm^{-1} could be the C=O of an ester. Check for two strong C—O stretches in the 1280–1050 cm^{-1} region.
- A stretch around 1700–1630 cm^{-1} might mean you have an amide. You can confirm this by checking for an N—H stretch back up the spectrum, but it will only show up for primary and secondary amides. Tertiary amides have no hydrogens on their N's.

Region: 1680–1400 cm^{-1}. The C=C stretch can show up here, the aliphatic C=C around 1680–1600 cm^{-1}, and the aromatic C=C around 1600–1400 cm^{-1}. If the C=C has any H's on it, you can confirm this if there's a C=C—H stretch up around 3000 cm^{-1}.

Region: 1500–600 cm^{-1}. The fingerprint region. A lot of the peaks here are due to the characteristics of the entire molecule and, just like fingerprints, they are unique to every molecule. That's not to say that there aren't characteristic functional group bands here. It's just that assigning them can be problematic.

Region: 850–500 cm^{-1}. The absorptions for halides are around here. The C—Cl stretch can show up almost anywhere in this area. The C—Br stretch is usually relegated to the area below 650.

Regions: 2000–1667 and 910–690 cm⁻¹. Aromatic ring substitution patterns. The aromatic C—H out-of-plane bending around 910–690 generates medium-to-strong peaks that can indicate the substitution pattern of an aromatic compound. The 2000–1667 region can also be used in this manner, but the peaks can be much weaker.

Did you notice? I stopped using the cm⁻¹ notation. While you don't do this when you're writing papers, or in your notebook, or even in your prelab exercises, few people add the word "wavenumbers" or the term "reciprocal centimeters" to the values when they are talking about them.

INFRARED SAMPLE PREPARATION

You can prepare samples for IR spectroscopy easily, but you must strictly adhere to one rule:

No water!

In case you didn't get that the first time:

No water!

Ordinarily, you put the sample between two sodium chloride plates. Yes. Common, ordinary *water-soluble* salt plates. Or mix it with **potassium bromide (KBr)**, another water-soluble salt.

So keep it dry, people.

Liquid Samples

1. Make sure the sample is DRY. NO WATER!

2. Put some of the *dry* sample (1–2 drops) on one plate, then cover it with another plate (Fig. 32.11). The sample should spread out to cover the entire plate. You *don't have to press*. If it doesn't cover well, try turning the top plate to spread the sample, or add more sample.

3. Place the sandwich in the IR salt plate holder and cover it with a hold-down plate.

4. Put at least two nuts on the posts of the holder (opposite corners) and spin them down *GENTLY* to hold the plates with an even pressure. *Do not use force!* You'll crack the plates! Remember, these are called salt plate holders and not salt plate smashers.

5. Slide the holder and plate into the bracket on the instrument in the sample beam (closer to you, facing the instrument).

6. Run the spectrum.

DO NOT OVERTIGHTEN

Two hex nuts
on diagonal posts

Top slides down
over screw posts

NaCl plate

One drop of sample
placed here

NaCl plate

Four screw posts

Back plate
fits into holder
on spectrophotometer

Sandwich of
two salt plates
with sample
between them

"V"– holds salt
in IR path

Back plate fits into holder
on spectrophotometer

FIGURE 32.11   IR salt plates and holders.

7. Take the salt cells out of the instrument and clean them. Often a little acetone followed by drying with a Kimwipe will suffice. Salt plates need a clean, *dry* place to rest in after work. Help them get home.

Since you don't have any other solvents in there, just your liquid compound, you have just prepared a liquid sample, **neat,** meaning *no solvent.* This is the same as a *neat liquid sample,* which is a way of describing any liquid without a solvent in it. It is not to be confused with a "really neat liquid sample," which is a way of expressing your true feelings about your sample.

Solid Samples

The Thin-Film Solid   A fast, simple method for getting the IR of a solid. Dissolve your solid in a small amount of a rapidly evaporating solvent—methylene chloride, diethyl ether, pentane—then put a drop of the solution on the plate and wait for the solvent to evaporate. Put the single, solid-coated cell in the salt plate holder, and run the IR as if this were a liquid between salt plates. Some say the salt plate should be gently warmed before dropping the solution on it. Check out P. L. Feist, *J. Chem. Educ.* **78,** 351 (2001) for all the gory details. Once you've tried this and seen it work, most other methods, including the fairly rapid Nujol mull, get left in the dust.

The Nujol Mull   A rapid, inexpensive way to get an IR of solids is to mix them with Nujol, a commercially available mineral oil. Traditionally, this is called "making a Nujol mull," and it is practically idiomatic among chemists. Although you won't see Rexall or Johnson & Johnson mulls, the generic name **mineral oil mull** is often used.

You want to disperse the solid throughout the oil, making the solid transparent enough to IR that the sample will give a usable spectrum. Since mineral oil is a saturated hydrocarbon, it has an IR spectrum all its own. You'll find hydrocarbon bends, stretches, and pushups in the spectrum, but you know where they are, and you ignore them. You can either look at a published reference Nujol spectrum (Fig. 32.12) or run your own if you're not sure where to look.

1. Put a small amount of your solid into a tiny agate mortar, and add a few drops of mineral oil.

2. Grind the oil and sample together until the solid is a fine powder *dispersed throughout the oil*.

3. Spread the mull on one salt plate, and cover it with another plate. There should be no air bubbles, just an even film of the solid in the oil.

4. Proceed as if this were a *liquid sample*.

5. Clean the plates with *anhydrous* acetone or ethanol. *NOT WATER!* If you don't have the tiny agate mortar and pestle, try a Witt spot plate and the rounded end of a thick glass rod. The spot plate is a piece of glazed porcelain with dimples in it. Use one as a tiny mortar and the glass rod as a tiny pestle.

And remember to forget the peaks from the Nujol itself.

Solid KBr Methods   KBr methods (hardly ever called potassium bromide methods) consist of making a mixture of your solid (*dry again*) with IR-quality KBr. Regular KBr off the shelf is likely to contain enough nitrate, as KNO_3, to give spurious peaks, so don't use it. After you have opened a container of KBr, dry it and later store it in an oven, with the cap off, at about 110°C to keep the moisture out.

Preparing the Solid Solution

1. At least once in your life, *weigh out* 100 mg of KBr so you'll know how much that is. If you can remember what 100 mg of KBr looks like, you won't have to weigh it out every time you need it for IR.

2. Weigh out 1–2 mg of your *dry, solid* sample. You'll have to weigh out *each sample* because different compounds take up different amounts of space.

3. Pregrind the KBr to a fine powder about the consistency of powdered sugar. Don't take forever, since moisture from the air will be coming in.

Pressing a KBr Disk—the Minipress (Fig. 32.13)

1. Get a clean, dry press and two bolts. Screw one of the bolts about halfway into the press and call that the bottom of the press.

2. Scrape a finely ground mixture of your compound (1–2 mg) and KBr (approximately 100 mg) into the press, so that an even layer covers the bottom bolt.

FIGURE 32.12 A published reference Nujol spectrum.

287

Turn "top" bolt down
to form pellet

Barrel

Sample on this bolt

"Bottom" bolt halfway in

FIGURE 32.13 The minipress.

3. Take the other bolt and turn it in from the top. *Gently* tighten and loosen this bolt at least once to spread the powder evenly on the face of the bottom bolt.

4. Hand-tighten the press again, and then use wrenches to tighten the bolts against each other. Don't use so much force that you turn the heads right off the bolts.

5. Remove both bolts. A KBr pellet containing your sample should be in the press. Transparent is excellent. Translucent will work. If the sample is opaque, you can run the IR, but I don't have much hope of your finding anything.

6. Put the entire press in a holder placed in the analyzing beam of the IR, as in Fig. 32.14. (Don't worry about that yet; I'll get to it in a moment. See "Running the Spectrum.")

Metal plate slips into
sample beam aperture

The minipress
(bolts removed,
of course)

KBr "window" with
sample in it

FIGURE 32.14 The minipress in its holder.

Pressing a KBr Disk—the Hydraulic Press   If you have a hydraulic press and two steel blocks available, there is another easy KBr method. It's a card trick, and at no time do my fingers leave my hands. The only real trick is that you'll have to bring the card.

1. Cut and trim an index card so that it fits into the **sample beam aperture** (see Figure 32.19).

2. Punch a hole in the card with a paper punch. The hole should be *centered in the sample beam* when the card is in the sample beam aperture.

3. Place one of the two steel blocks on the bottom jaw of the press.

4. Put the card on the block.

5. Scrape your KBr–sample mixture onto the card, covering the hole and some of the card. Spread it out evenly.

6. Cover the card, which now has your sample on it, with the second metal block and place the sandwich in the press (Fig. 32.15).

7. Pump up the press to the indicated *safe* pressure.

8. Let this sit for a bit (1 minute). If the pressure has dropped, bring it back up, *slowly, carefully,* to the safe pressure line, and wait another bit (1 minute).

9. Release the pressure on the press. Push the jaws apart.

10. Open the metal sandwich. Inside will be a file card with your sample in a **KBr window,** just like in the minipress.

    Caution! The KBr window you form is rather fragile, so don't beat on it.

11. Put the card in the **sample beam aperture** (Fig. 32.16). The KBr window should be *centered* in the sample beam if you've cut and punched the card correctly.

FIGURE 32.15   KBr disks by hydraulic press.

Sample holder plate slips
down into groove

Reference aperture

Sample aperture

FIGURE 32.16   Putting sample holders with samples into the beam.

RUNNING THE SPECTRUM

There are many IR instruments, and since they are so different, you need your instructor's help here more than ever. There are a few things you have to know.

1. *The sample beam.* Most IRs are **dual-beam instruments** (Fig. 32.17). The one closest to you, if you're operating the instrument, is the **sample beam.** Logically, there is a **sample holder** for the sample beam, and your sample goes there. And then a beam of IR radiation goes through your sample.

2. *The reference beam.* This is the other light path. It's not *visible* light but another part of the electromagnetic spectrum. Just remember that the reference beam is the one farthest away from you.

3. *The 100% control.* This sets the pen at the 100% line on the chart paper. Or tries to. It's a very delicate control and doesn't take kindly to excessive force. Read on, and all will be made clear.

FIGURE 32.17 Schematic diagram of an IR.

4. ***The pen.*** There is a pen and pen holder assembly on the instrument. This is the way the spectrum is recorded on the chart paper. Many people get the urge to throw the instrument out the window when the pen stops writing in the middle of the spectrum. Or doesn't even start writing. Or was left empty by the last fellow. Or was left to dry out on the top of the instrument. For those with Perkin-Elmer 137s and 710s and the like, two clever fellows have made up generic felt-tip pen holders for the instruments. This way, you buy your own pen from the bookstore, and if it dries out it's your own fault. [See R. A. Bailey and J. W. Zubrick, *J. Chem. Educ.*, **59,** 21 (1982).] Unfortunately, the number of truly skinny pens has declined in recent years, and these adapters might be less useful.

5. ***The very fast or manual scan.*** To get a good IR, you'll have to be able to scan, very rapidly, *without letting the pen write on the paper.* This is so you'll be able to make adjustments before you commit pen to paper. This **fast forward** is not a standard thing. Sometimes you operate the instrument by hand, pushing or rotating the paper holder. Again, *do not use force.*

Whatever you do, *don't try to move the paper carrier by hand when the instrument is scanning a spectrum.* Stripped gears is a crude approximation of what happens. So, thumbs off.

I'm going to apply these things in the next section using a real, albeit no longer produced instrument, the Perkin-Elmer 710B, as a model. Just because you have another model is no reason to skip this section. If you do have a different IR, try to find the similarities between it and the Perkin-Elmer model. Ask your instructor to explain any differences.

THE PERKIN-ELMER 710B IR (FIG. 32.18)

1. *On–off switch and indicator.* Press this once, the instrument comes on, and the switch lights up. Press this again, the instrument goes off, and the light goes off.

2. *Speed selector.* Selects speed (normal or fast). "Fast" is faster, but slower gives *higher resolution,* that is, more detail.

3. *Scan control.* Press this to *start a scan.* When the instrument is scanning, the optics and paper carrier move automatically, causing the IR spectrum to be drawn on the *chart paper* by the *pen.*

4. *Chart paper carriage.* This is where the chart paper nestles while you run an IR. If it looks suspiciously like a clipboard, it's because that's how it works.

5. *Chart paper hold-down clip.* Just like a clipboard, this holds the paper down in the carrier.

6. *Frequency scale.* This scale is used to help align the chart paper and to tell you where in the spectrum the instrument is during the run.

7. *Scan position indicator.* A white arrow that points, roughly, to where the instrument is in the spectrum.

FIGURE 32.18   The Perkin-Elmer 710B IR.

8. *Line-up mark.* A line, here at the number 4000, that you use to match up the numbers on the instrument frequency scale with the same numbers on the chart paper.

9. *Pen and transmittance scale.* This is where the pen traces your IR spectrum. The numbers here mean percentage of IR transmitted through your sample. If you have no sample in the sample beam, how much of the light is getting through? Those who said 100% are 100% correct. Block the beam with your hand and 0% gets through. You should be able to see why these figures are called the **percent transmission,** or **%T,** scale.

10. *The 100% control.* Sets the pen at 100%. Or tries to. This is a fairly sensitive control, so don't force it.

11. *Sample beam aperture.* This is where you put the holder containing your sample, be it mull or KBr pellet. You slip the holder into the aperture window for analysis.

12. *Reference beam aperture.* This is where nothing goes. Or, in extreme cases, you use a **reference beam attenuator** to cut down the amount of light reaching the detector.

USING THE PERKIN-ELMER 710B

1. Turn the instrument on, and let it warm up for about 3–5 minutes. Other instruments may take longer.

2. Get a piece of IR paper and load the **chart paper carriage,** just like a clipboard. Move the paper to line up the index line on the paper with the index line on the instrument. It's at 4000 cm^{-1}, and it's only a rough guide. Later, I'll tell you how to calibrate your chart paper.

3. Make sure the chart paper carriage is at the high end of the spectrum, 4000 cm^{-1}.

4. Put your sample in the sample beam. Slide the sample holder with your sample into the sample beam aperture (Fig. 32.16).

5. What to do next varies for particular cases—not by much, but enough to be confusing in setting things up.

6. Look at where the pen is. *Carefully* use the **100% control** to locate the pen at about the 90% mark when the chart (and spectrum) is at the high end, 4000 cm^{-1}.

The 100% Control: An Important Aside

Usually, there's not much more to adjusting the 100% control than is shown in upcoming items 7 through 9, unless your sample, *by its size alone,* reduces the amount of IR that reaches the detector. This really shows up if you've used the minipress, which has a *much smaller opening* than the opening in the reference beam.

Gently slide reference
beam attenuator and
watch pen — stop when
pen points to 80–90% T

KBr in card (or minipress)
blocks sample beam and
must be compensated for

FIGURE 32.19 Using a reference beam attenuator with a KBr window.

So you're at a disadvantage right from the start. Sometimes the 100% control mopes around at much less than 40%. That's terrible, and you'll have to use a **reference beam attenuator** (Fig. 32.19) to equalize the amounts of energy in the two beams. As you block more and more of the energy in the reference beam, the %T will go back to the 100% mark. Stop at about 90%. Note that, if you didn't have these problems, this is where you'd put the pen with the 100% control anyway. Use the *smallest amount* of reference beam attenuation you can get away with.

7. OK, at 4000 cm^{-1} the %T (the pen) is at 90%.

8. Now *slowly, carefully,* move the pen carriage manually so that the instrument scans the entire spectrum. *Watch the pen!* If the **baseline** creeps up and goes off the paper (Fig. 32.20), this is not good. Readjust the 100% control to keep the pen on the paper. Now, keep going, slowly, and *if* the pen drifts up again, readjust it again with the 100% control and get the pen back on the paper.

9. Now, go back to the beginning (4000 cm^{-1}). If you've adjusted the 100% control to get the pen back on the paper at some other part of the spectrum, surprise! The pen will not be at 90% when you get back. This is unimportant.

FIGURE 32.20   An IR with an unruly baseline.

What is important is that *the pen stays within the limits* between the *0 and 100%T lines for the entire spectrum.*

10. In any case, if the peaks are too large and the baseline is in the proper place, your sample is just *too concentrated.* You can wipe some of your liquid sample or mull off one of the salt plates or remake the KBr pellet using less compound or more KBr. Sorry.

11. When you've made all the adjustments, press the scan button and you're off.

CALIBRATION OF THE SPECTRUM

Once the run is over, there's one other thing to do. Remember that the index mark on the paper is not exact. You have to **calibrate the paper** with a standard, usually *polystyrene film.* Some of the peaks in polystyrene are quite sharp, and many of them are very well characterized. A popular one is an extremely narrow, very sharp spike at 1601.4 cm^{-1} (6.24 μm).

1. *Don't move the chart paper,* or this calibration will be worthless.

2. Remove your sample, and replace it with the standard polystyrene film sample. You will have to remove any reference beam attenuator and turn the 100% control to set the pen at about 90% when the chart is at 4000 cm^{-1}.

3. *With the pen off the paper,* move the carriage so that it's *just before* the calibration peak you want, in this case 6.24 μm.

Micrometers (μm)

Fixes
locations
of all
other peaks

Polystyrene
1601.4 cm^{-1}

Frequency (cm^{-1})

FIGURE 32.21 A calibration peak on an IR spectrum.

4. Now, quickly start the scan, *let the pen draw just the tip of the calibration peak,* and quickly stop the scan. You don't need to draw more than that (Fig. 32.21). Just make sure you can pick your calibration peak out of the spectral peaks. If it's too crowded at 1601.4 cm^{-1}, use a different polystyrene peak—2850.7 or 1028.0 cm^{-1} (3.51 or 9.73 μm). Anything really well known and fairly sharp will do (Fig. 32.22).

5. And that's it. You have a nice spectrum.

IR SPECTRA: THE FINISHING TOUCHES (FIG. 32.23)

IR chart paper contains spaces for all sorts of information. It would be nice if you could fill in:

1. ***Operator.*** The person who ran the spectrum. Usually you.

2. ***Sample.*** The name of the compound you've just run.

3. ***Date.*** The day you ran the sample.

4. ***Phase.*** For KBr, say "solid KBr." A Nujol mull is "Nujol mull." Liquids are either solutions in solvents or "neat liquids," that is, without any solvents, so call them liquids.

5. ***Concentration.*** For KBr, a solid solution, list milligrams of sample in 100 mg of KBr. For liquids, *neat* is used for liquids without solvents.

FIGURE 32.22   IR of polystyrene film, pointing out many calibration peaks.

6. *Thickness.* Unless you're using solution cells, write **thin film** for **neat liquids.** Leave this blank for KBr samples (unless you've measured the thickness of the KBr pellet, which you shouldn't have done).

7. *Remarks.* Tell where you put your calibration peak, where the sample came from, and anything unusual that someone in another lab might have trouble with when trying to duplicate your work. Don't put this off until the last day of the semester when you can no longer remember the details. Keep a record of the spectrum *in your notebook.*

You now have a perfect IR, suitable for framing and interpreting.

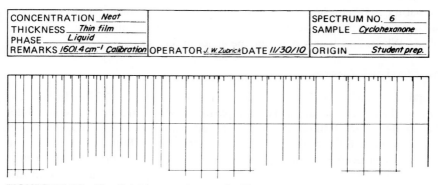

FIGURE 32.23   The finishing touches on the IR.

INTERPRETING IR SPECTRA— FINISHING TOUCHES

IR interpretation—finishing touches can be as simple or as complicated as you'd like to make it. You've already seen how to distinguish alcohols from ketones by **correlation** of the positions and intensities of various peaks in your spectrum with positions listed in **IR tables** or **correlation tables.** This is a fairly standard procedure and is probably covered very well in your textbook. The things that are not in your text are as follows:

1. *Not forgetting the Nujol peaks.* Mineral oil gives huge absorptions from all of the C—H bonds. They'll be the biggest peaks in the spectrum. And every so often, people mistake one of these for something that belongs with the sample.

2. *Nitpicking a spectrum.* Don't try to interpret every wiggle. There is a lot of information in an IR, but sometimes it is confusing. Think about what you're trying to show, and then show it.

3. *Negative results.* Negative results can be very useful. No big peak between 1600 and 2000 cm^{-1}? No carbonyl. Period. No exceptions.

4. *Pigheadedness in interpretation.* Usually a case of "I know what this peak is, so don't confuse me with facts." Infrared is an extremely powerful technique, but there are limitations. You don't have to go hog wild over your IR, though. I know of someone who decided that a small peak was an N—H stretch, and the compound *had* to have nitrogen in it. The facts that the intensity and position of the peak were not quite right and neither a chemical test nor solubility studies indicated nitrogen didn't matter. Oh well.

THE FOURIER TRANSFORM INFRARED (FTIR)

There's even more of a difference in particulars between FTIRs than between dispersive IRs (like the 710B mentioned earlier) because not only can the basic instrument vary, but it can be attached to any number of different computers of differing configurations running any number of different software programs doing the infrared analysis. Whew. So I'll only highlight the major differences.

The Optical System

The optical system is based on the Michelson interferometer (Fig. 32.24). Infrared energy from the source goes through the sample in a single beam and hits a beam splitter in the interferometer. Half of the light is reflected to a stationary mirror, and the other half passes through the beam splitter to a moving mirror. Both mirrors reflect the light back to the beam splitter, where they recombine to form an interference pattern of constructive and destructive interferences known as an **interferogram.** (Consider

Stationary mirror

Moving mirror

Infrared source

Sample

Beam splitter

Infrared detector

FIGURE 32.24 Michelson interferometer optical system for FTIR.

the pattern produced when sets of waves from two stones thrown into a pond interact: The peak of one and the peak of the other reinforce each other, giving a bigger peak; the peak of one and the trough of the other can completely destroy each other and produce no displacement.) That interference pattern varies with the displacement of the moving mirror, and this pattern of variation is sent along to the detector.

A computer programmed with the algorithm of the Fourier transformation converts the measured-intensity-versus-mirror-displacement signal (the interferogram) into a plot of intensity versus frequency—our friendly infrared spectrum.

FTIR has a few advantages:

1. *Fellgett's advantage (multiplex advantage).* The interferometer doesn't separate light into individual frequencies like a prism or grating, so every point in the interferogram (spectrum) contains information from each wavelength of infrared radiation from the source.

2. *Jacquinot's advantage (throughput advantage).* The dispersive instrument, with prism or grating, needs slits and other optics so as much as possible of a single wavelength of energy reaches the detector. Not so the interferometer. The entire energy of the source comes roaring through the optical system of the interferometer to the detector, increasing the signal-to-noise ratio of the spectrum.

3. *Conne's advantage (frequency precision).* The dispersive instrument depends on calibration (polystyrene at 1601 cm^{-1}) and the ability of gears and levers to move slits and gratings reproducibly. The FTIR carries its own internal frequency standard, usually a He–Ne gas laser, which serves as the master timing clock and tracks mirror movement and frequency calibration to a precision and accuracy of better than 0.01 wavenumbers (cm^{-1}).

FIGURE 32.25   FTIR background spectrum. Note H_2O and CO_2 bands.

But what about your advantages and your disadvantages?

1. ***FTIR is a single-beam instrument.*** So you must collect a background spectrum (Fig. 32.25) before you do your sample. The computer program will subtract the background from the background and sample and produce the usual IR spectrum. What's in a background spectrum? Everything that's not the sample. For a KBr disk spectrum, the empty press (Fig. 32.14) or an empty card (Fig. 32.15) is included in the background spectrum. For a salt plate spectrum (Fig. 32.11), clean salt plates in the holder are the background. And for a solution-cell spectrum, the cell with pure solvent is the background. Why? For many cells and holder combinations, the aperture is smaller, or the salt plates absorb a bit, or the solvent absorbs a lot, and these must be compensated for.

2. ***The FTIR is fast.*** I can get the spectrum of a compound on my computer screen in about 0.21 minute, considerably faster than any dispersive instrument. Unfortunately, our inexpensive plotter is a bit slow, and, as the software makes the plotter label the axes with numbers and text, it does take a bit more than the 0.21 minute. Overall, though, it is really considerably faster.

3. ***The FTIR is accurate.*** So you don't have to run a reference polystyrene spectrum. Unfortunately, we plot the spectra on inexpensive copier paper, and if you don't mark relevant peaks on the screen with the software, you may never get those really accurate frequencies to 0.01 cm^{-1}.

4. ***The FTIR computer programs automatically scale the spectrum.*** So you don't have to set the zero and 100%. Unfortunately, the scaling is mindless, and you

FIGURE 32.26   Salt plate spectrum transmittance trick.

can be fooled if you're not careful. Figure 32.26 could possibly be an organic acid. See the broad hydrogen-bonded OH peak at about 3400 cm^{-1}? (The consequences of not marking the exact frequency with the computer program should be evident now.) But the C=O at 1750 cm^{-1} isn't as strong as it should be and . . . Fooled ya! Look at the transmittance scale on the left-hand side. It only goes from 70% to 76%. This is a spectrum of some student-abused salt plates, and these huge peaks and bands are actually tiny wiggles (Fig. 32.27). The computer program mindlessly took the maximum and minimum wiggling in the spectrum and decided to give it to you at full-screen height.

Admittedly, this is an unusual case. Since my background spectrum was taken with an empty salt plate holder in the beam, my salt plates became the spectrum. Had I run the background with the holder and the cells, any bumps from the salt plates would have been canceled. And to get that 0-to-75% transmittance spectrum, I had to go outside the usual operating range of our instrument. Normally, you won't have such problems. But if you are now on your guard, you won't have *these* problems.

5. *The FTIR programs usually have a library spectral search, so you can easily identify your compounds.* Unfortunately, not every sample you'll run is in the database. So you could wind up with a collection of compounds that usually have the same functional groups (and that can be helpful). Close, but no cigar. In that case, you must not succumb to the "I got it off the computer, it must be right" syndrome.

FIGURE 32.27 Salt plate spectrum with some sanity restored.

AND UV-VIS TOO!

Only, and I have to stress this, *only* because the instrumentation and sample handling have some similarities is this section here. For instance, many UV-VIS (that's how to speak about them) instruments, just like their infrared counterparts, are dual-beam, have holders for both a sample cell and a reference cell, and work in the same way. And, because of computers and CCDs (charge coupled devices), single-beam UV-VIS instruments also exist.

ELECTRONS GET TO JUMP

In IR spectrometry, it's the bending and stretching of the bonds between atoms that give rise to the spectrum. In UV-VIS, outer electrons are excited directly, and it's these electronic transitions that give rise to the spectrum (Fig. 32.28).

Electrons in π and σ bonds, as well as nonbonding (n) electrons, can be involved.

Most of the UV-VIS spectroscopy of organic compounds involves $n \rightarrow \pi^*$ and $\pi \rightarrow \pi^*$ transitions that happen to fall in the 200–700-nm range of UV-VIS spectrophotometers.

FIGURE 32.28 Possible electronic transitions.

INSTRUMENT CONFIGURATION

The dual-beam configuration of UV-VIS instrumentation does have a lot of similarities to that of the IR (Fig. 32.29). In the modern single-beam instrument, a CCD or diode array sees the entire spectrum and sends that information to the computer (Fig. 32.30).

FIGURE 32.29 Dual-beam spectrometer.

FIGURE 32.30 Modern charge-coupled-device single-beam spectrometer.

Source

Traditionally, what amounts to a lightbulb produced the visible spectrum, and a low-pressure hydrogen lamp produced the UV spectrum. Quartz-halogen lamps now may provide a single source for both regions. In all cases, when replacing these lamps, don't leave fingerprints on the glass or quartz.

Sample (and Reference) Cells

VIS-UV cells are square cells 1.0 cm on a side, and can be made of quartz glass, flint glass, or plastic. Only the quartz cells can pass UV as well as visible radiation, so a single set of these would be the most versatile. Plastic cells, while inexpensive, not only don't pass UV, they can dissolve in certain solvents.

Superficially, glass and quartz cells look the same. But if you hold them up to light, and look down the top edges of the cells, quartz cells are bright white, while glass cells tend to be greener. Set up the spectrometer to sit at a UV wavelength (I suggest 254 nm, the same UV wavelength you'd use to light up the green fluorescent substance in TLC plates), and if you get very little or no UV coming through the cells, they're glass.

In a dual-beam spectrometer, consider getting pairs of matched cells. This way you eliminate one more variable in taking spectra. And since they are square cells, they have four sides, with two opposing sides clear and the other two sides frosted (glass, quartz) or rippled (plastic). It can be frustrating, and potentially embarrass-ing, to find out that you're having trouble getting a good spectrum because the beam is traveling through the frosted sides. Give the cells a one-quarter turn, and start again.

Solvents

Solvents should not be opaque to the regions you're examining with the UV-VIS spectrometer. You can't necessarily use the ubiquitous organic laboratory solvents you've used for reactions, extractions, and recrystallizations. They may have traces of impurities with their own absorptions that can interfere with the absorption spec-trum of your sample. You have to use very pure solvents. Because most HPLC units have ultraviolet detectors, HPLC solvents should be considered for your work.

EXERCISES

1. If a bond stretches in the forest, and the dipole moment doesn't change, does it make an absorption?
2. If you prepare a KBr pellet, and you can't see through it, can the IR?

3. A sample gives strong peaks at both 3300 and 1750. What kind of compound might it be? Would it make a difference whether the peak at 3300 is described as very broad or just broad?

4. Someone said that IRs of aromatic compounds are "more spiky" than those of the corresponding aliphatics. Recast this description in less colorful, more boring, but perhaps more accurately detailed terms.

NUCLEAR MAGNETIC RESONANCE

Nuclear magnetic resonance (NMR) can be used like IR to help identify samples. But if you thought the instrumentation for IR was complicated, these NMR instruments are even worse. NMR even has a Fourier transform variant called, you got it, FT-NMR. So I'll only give some generalities and directions for preparing samples.

For organic lab, traditionally you look only at the signals from protons in your compound; sometimes, then, this technique is called *proton* **magnetic resonance** (**¹H NMR**). Surely I don't mean naked H¹ protons, do I? No. The everyday **hydrogens** in organic compounds are just called **protons** when you use this technique.

A sample in a special tube is spun about its major axis between the poles of a strong magnet. A radio-frequency signal, commonly 60 MHz, a little higher than what used to be TV Channel 2, is applied to the sample. You get better resolution of proton signals at higher frequencies, but the instruments, and even the NMR tubes, cost more.

Anyway, if *all* protons were in the same environment, there'd be this big absorption of energy in one place in the ¹H NMR spectrum. Big deal. But all protons are *not* the same. If they're closer to electronegative groups or on aromatic rings, the signals shift to a different frequency. This change in the position of the ¹H NMR signals, which depends on the chemistry of the molecule, is called the **chemical shift.** Thus, you can tell quite a bit about a compound if you have its NMR (i.e., its NMR spectrum). In effect, NMR is how we listen to what protons are saying about themselves and about their neighbors!

NUCLEI HAVE SPIN, TOO

Actually, different nuclei can have different spin states: half-integer spin (like those electrons), integer spin, and no spin. The letter I represents these total nuclear spin angular momentum quantum numbers.

If the number of neutrons and the number of protons are both even, the nucleus has no spin. Both the carbon-12 nucleus, with 6 protons and 6 neutrons, and the oxygen-16 nucleus, with 8 protons and 8 neutrons, have zero spin and are NMR-inactive.

If the number of neutrons plus the number of protons is odd, then the nucleus has a half-integer spin quantum number (1/2, 3/2, 5/2, . . .). Protons (hydrogen-1) have a spin of 1/2, and from the selection rule $2I + 1$ have two states, $+1/2$ and $-1/2$. But there's a catch.

THE MAGNETIC CATCH

Unless an external magnetic field is applied, the two nuclear spin states of protons are said to be **degenerate.** This means that the two states are equal in energy: With an externally applied magnetic field, the energy levels separate, with one state (α) lower in energy, and the other (β) higher in energy. And in the NMR experiment, the point is to supply energy so that the protons can absorb that energy and change, or flip, from the lower to the higher state and then return to the lower state, a process known as **relaxation.**

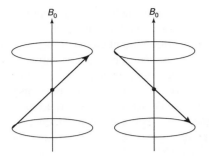

FIGURE 33.1 Processing protons sort of aligned with the external magnetic field in the low-energy and then high-energy state.

EVERYBODY LINE UP, FLIP, AND RELAX

So in this externally applied magnetic field, B_0, the protons are distributed between these two energy states either sort of with the field or sort of against the field, as in Figure 33.1. The spinning nucleus takes on a precessional orbit, much like a spinning top, at a frequency (v) called the Larmor frequency, determined by:

$$v = \gamma B_0$$

where B_0 is the magnetic field strength in tesla, and γ is the gyromagnetic constant (or magnetogyric constant) that's different for each nucleus. For protons, $\gamma = 4.26 \times 10^7$ Hz/T, where Hz is the frequency in hertz (formerly cycles per second, with units $\sec^{-1}$) and T is tesla, the magnetic field strength. What's important here is to note that the higher the field strength, the higher the resonant frequency.

Early NMR instruments had a base frequency of 60 megahertz (MHz) using a magnet with a field strength of about 1.4 T. More modern 300-MHz instruments use a 7.05-T magnet to get the protons in resonance. If we sweep either the magnetic field strength or the radio frequency energy so that we go through a proton's goldilocks point (where the magnetic field and the radio frequency are "just right"), the protons will absorb that energy and make the transition to the higher-energy state. Then, as we pass by that goldilocks point, the protons lose that energy by a process called relaxation. In the NMR instrument, there's a coil of wire that detects the energy change involved in this process, and it's that signal that gets amplified and sent to a computer to be recorded.

A MORE SENSITIVE CENSUS

One of the reasons for using both 60-MHz and 300-MHz resonant (Larmor) frequencies is to show not only that is there a difference in energy between the α and β states, but also that the difference in those frequencies makes a real difference in the NMR.

Since we have the frequencies, we can get the energies by using:

$$E = hv$$

For protons at 300 MHz, ΔE is about 0.029 cal/mole; for protons at 60 MHz it's about 0.0058 cal/mole. Now you can get the ratio of the population of protons in the β state, N_β, to that in the α state, N_α, from the Boltzmann equation:

$$\frac{N_\beta}{N_\alpha} = e^{-\Delta E/kT}$$

Using the ΔE values given, 19.85 cal/K · mole for the Boltzmann constant, k, and 300 K for the temperature, the ratio is 0.999999 at 60 MHz and 0.999995 at 300 MHz. Not the biggest of differences, but because fewer protons are in the higher-energy state at 300 MHz and 300 K, we get a stronger signal out. Thus, the NMR technique is more sensitive at the higher frequencies.

THE CHEMICAL SHIFT

If all the protons had the same Larmor frequency—resonated at the same frequency— then there would just be one absorption in the NMR for all compounds. Big deal. But that resonant frequency is determined by the magnetic field **at the nucleus.** So, if the magnetic field is different at the nucleus, the goldilocks point will be different for nuclei in different chemical environments, and resonate at different frequencies away from this basic frequency. So when you sweep through a range of Larmor (resonance) frequencies with the instrument, protons in different chemical environments will resonate at these different frequencies and show up at different places in the NMR spectrum.

How small are these differences? Pretty small. The chemical-shift units are in delta (δ), given in parts per million. Parts per million of what? Of the base frequency. So if an NMR signal falls at 1.0 δ, it is at a frequency only 60 Hz from the base in a 60-MHz instrument, or only 300 Hz from the base frequency in a 300-MHz instrument. This should also give you some insight as to why, early on, with only 60-MHz instruments available, NMR spectroscopists would say the chloroform peak is at 432 Hz rather than 7.1 δ. (Actually, they'd have said 432 cps.)

T FOR ONE AND TWO

During the spectrum sweep, protons in different magnetic environments will absorb energy as the sweep goes through the goldilocks point for various protons. When the sweep passes that point for a proton, it can't absorb the exciting energy anymore and starts to lose that energy, to relax. If the protons didn't fall back from their excited states and emit radio-frequency radiation, there would be no energy for the detection coil in the NMR to pick up, and no signal. There are two relaxation processes in NMR:

1. **Spin-lattice relaxation.** The entire NMR sample in the tube is said to be the lattice. All the nuclei in the lattice are in vibrational and rotational motion, making for a very complex magnetic field. Some of the components of the

lattice can interact with the protons in the higher-energy state and cause them to lose energy, increasing the rotational and vibrational energy of the entire lattice and resulting in a tiny temperature rise of the entire sample. The time it takes for nuclei to return to the lower-energy state is called T_1.

2. **Spin-spin relaxation.** In this case, a proton in the higher-energy state interacts with a neighboring proton in the lower-energy state, and they exchange quantum states. Thus, the just-excited-and-passed-by proton loses energy as a different proton gains that energy. The time it takes for the proton to lose that energy is called T_2.

BE IT BETTER RESOLVED . . .

Sure, at 300 MHz the energy difference between the α and β is larger than at 60 MHz, and sure, that makes the difference in the population of the two states greater at 300 MHz and smaller at 60 MHz, and sure, that makes the technique more sensitive at 300 MHz than at 60 MHz.

But running at the higher base frequency increases the resolution as well as the sensitivity. As much as we want to get away from using frequencies (Hz) in NMR and instead use δ (parts per million), Hz is always with us. Say there are two NMR signals, one at 3.0 δ and another at 4.0 δ. *They're separated by 1.0 δ no matter what the base frequency of the instrument is.* So, on a 60-MHz instrument they are separated by 60 Hz, and on the 300-MHz instrument their separation is five times greater because there are 300 Hz—not 60 Hz—in the separation. That could make for the difference between reporting an unresolved multiplet and reporting a discernible and analyzable pattern of NMR peaks.

INCREDIBLY BASIC FT-NMR

In the FT version of NMR, you excite all the protons at once, kill the pulse, and record the free induction decay (FID) signal. At that point, your computer takes over and transforms the data from the time domain into the frequency domain. Domain here means x-axis (the domain, remember; y-axis is called the range, eh?), so you eventually get the kind of NMR spectrum you've learned to know and love.

So how do you excite all the protons over the typical range of 10 δ all at once? Lots and lots of teenie NMR transmitters, perhaps? Perhaps it would be best to just blast the sample with a pulse, in which case, if you were paying attention in physics class, or reading articles on the frequency response of audio amplifiers in old issues of *Popular Electronics* instead of going to physics class, you'd know that a square wave can be made up of a series of sine waves of different frequencies. So the pulse itself can excite a range of protons all at once. Then, as the protons relax, the FID is stored away until the next pulse comes along. That FID is also stored away and signal-averaged.

When you've collected enough FIDs, the computer then transforms the averaged FID signal into the usual amplitude vs. frequency (in δ parts per million).

LIQUID SAMPLE PREPARATION

To prepare a liquid sample for NMR analysis:

1. Get an **NMR tube.** These tubes are about 180 mm long and 5 mm wide, and the cost is about three bucks apiece for what is euphemistically called the inexpensive model. They also have matching, color-coordinated designer caps (Fig. 33.2).

Disposable pipet and rubber bulb

1. Touch lightly to side—let sample go down the side of the tube. NMR tubes are fragile; don't snap the end.

3. Add 1 drop internal standard, then cap the tube with the NMR tube cap.

2. Fill to at least $\frac{3}{4}$ in.

FIGURE 33.2 Loading the typical NMR tube.

Don't pick up and hold the tube by its color-coordinated designer cap. The tube, perhaps with your sample in it, could come out of the cap and fall into the instrument or onto the floor. Hold NMR tubes *just below the cap* by the tube itself.

These inexpensive tubes are not precision-ground, and some may stick in the NMR probe. Usually, you don't get anywhere near the instrument; your instructor gets to run them, or give them to the guy who runs them, and you get a spectrum back. In the cases where you do get to insert a sample into the magnet yourself, the first thing you'll have to do is wipe the people grease off the outside of the tube with a Kimwipe or soft, clean cloth. Next, you put a plastic spinner on the tube, and set the height of the spinner from the bottom of the tube in a **depth gauge** on or near the magnet. If the spinner seems too tight or too loose, *you immediately stop and get your instructor to check the fit.* If the tube is too loose in the spinner, it may slip in the magnet sample holder and be difficult to remove without breaking something or having to take something apart. If the tube is too tight in the spinner, besides possibly damaging the spinner or breaking the tube as you foolishly continue to force it, it might not fit the magnet sample holder either and wind up broken or at some weird position in the chamber around the magnet, and might have to be extricated.

2. Get a **disposable pipet** and a little **rubber bulb,** and construct a **narrow medicine dropper.** Use this to transfer your sample to the NMR tube. Don't fill it much higher than about 3–4 cm. Without any solvent, this is called, of course, a **neat sample.**

3. Ask about an **internal standard.** Usually **tetramethysilane (TMS)** is chosen because most other proton signals from any sample you might have fall at *lower frequencies* than that of the protons in TMS. Sometimes **hexamethyldisiloxane (HMDS)** is used because it doesn't boil out of the NMR tube like TMS can. TMS boils at 26–28°C; HMDS boils at 101°C. Add *only* one or two drops of internal standard.

4. Cap the tube and have the NMR of the sample taken. It's really out of place for me to tell you more about NMR here.

5. Last point: cleanliness. If there is *trash* in the sample, get rid of it. Filter it or something, will you?

6. Speaking of cleanliness, NMR tubes can be cleaned with a professionally available NMR tube cleaner, or pipe cleaners. Yep, pipe cleaners. Now I seem to recall you could get a roll of this stuff—perhaps that's that twist-tie material people use to tie their tomato plants up with—but if you can't, you can carefully twist together two pipe cleaners at the ends to make one cleaner long enough to do the job. You have to be a little careful putting the twisted section into the tube, but it does work. Leave the tube open to dry . . . and don't even think of blowing unfiltered, dirt- and oil-laden air from the lab air jets into the tube to dry it.

SOLID SAMPLES

Lucky you. You have a solid instead of a liquid. This presents one problem: What are you going to dissolve the solid in? Once it's a solution, you handle it just like a **liquid sample.** Unfortunately, if the *solvent has protons* and you know there'll be much more solvent than sample, you'll get a major proton signal from your solvent. Not a good thing, especially if the signals from your solvent and sample overlap.

Protonless Solvents

Carbon tetrachloride, a solvent without protons, is a typical NMR solvent. In fact, it's practically the only protonless solvent. So if your sample dissolves in CCl_4, you're golden. Get at least 100 mg of your compound in enough solvent to fill the NMR tube to the proper height.

CAUTION! CCl_4 is toxic and potentially carcinogenic. Handle with extreme care.

You might not be able to use CCl_4 anyway. Most NMR instruments have an entire, but much simpler, second NMR instrument inside them made specifically to lock onto the NMR signal of deuterium in the solvent in your sample. This **deuterium lock signal** is used to compensate for any drift in the magnet or electronics as your protons spectrum gets swept. While you can collect a spectrum without the lock, it might be of a lesser quality. And if you change your deuterated solvent from deuterated chloroform to, say, deuterated acetone, you might have to make an adjustment to this second unit to maintain the deuterium lock.

Deuterated Solvents

If your compound does not happen to dissolve in CCl_4, you still have a shot because *deuterium atoms do not give ^1H NMR signals.* This is logical, since they're not protons. The problem is that *deuterated solvents are expensive,* so do NOT ask for, say, D_2O or $CDCl_3$, the **deuterated analogs of water** and **chloroform,** unless you're absolutely sure your compound will dissolve in them. Always use the protonic solvent—H_2O or $CHCl_3$ here—for the solubility test. There are other deuterated solvents that may or may not be available for use. Check with your instructor.

FT-NMR SAMPLE PREPARATION

Unlike continuous-wave NMR, for which the sample should be as concentrated as possible, FT-NMR samples are diluted. We're talking 3–5% for both solids and liquids, so you can't really have neat liquid samples anymore. The solvent you use has to have two properties: It has to **dissolve** your sample, and it has to be **deuterated,** not just *sans* protons, but **deuterated.** That deuterium lock thing again.

The best FT-NMR spectra are taken with multiple sweeps of the sample spectrum. The multiple sweeps are subjected to signal averaging, and, without putting too fine a point on it, noise, being random, tends to average to zero, while signal, being signal, tends to add up. Thus, your spectrum improves. Unless . . . unless the magnetic field is unstable and wanders as you collect the sweeps, in which case the signal and the noise shift from sweep to sweep and you get noise adding to signal and signal adding to noise and a mess.

To keep the spectrum in place, the instrument uses a **deuterium lock,** a signal from the solvent that helps to keep the spectrum in place. That's why you need the right solvent, *deuterated*!

In addition, with a sample that's—what?—95% solvent, the NMR signal from protons in the solvent would overwhelm any signal from the sample.

To prepare a sample for FT-NMR:

1. ***Find the solvent for your sample.*** Deuterated solvents are fairly expensive, so you'll probably have a short list. Deuterated water and deuterated chloroform are very common; you might have others. To check the solubility, *do not use the deuterated solvents! They're expensive.* Use the common proton variants for your testing. Only when you are sure that the solvent you are going to use will dissolve your sample, use the deuterated solvent.

2. ***Weigh some things.*** Weigh your empty NMR tube. Put an inch or so of the solvent into the tube, and weigh the tube again. Subtract, and get the mass of the solvent. Find 3–5% of that number and you have the mass of sample, liquid or solid, that you should put into the tube.

3. ***Add the standard.*** Add 1 drop of TMS or HMDS to act as the internal standard.

Now you can run the sample in the FT-NMR.

SOME NMR INTERPRETATION

I've included a spectrum of ethylbenzene (Fig. 33.3) to give you some idea of how to start interpreting NMRs. Obviously, you'll need more than this. See your instructor or any good organic chemistry text for more information.

The Zero Point

Look at the extreme right of the NMR spectrum. That single, sharp peak comes from the protons in the **internal standard,** TMS. This signal is *defined as zero,* and all other values for the **chemical shift** are taken from this point. Again, the units are parts per million (ppm) and use the Greek letter delta, δ: δ 0.0.

Protons of almost all other compounds you'll see will give signals *to the left of zero:* positive values, **shifted downfield** from TMS. There are compounds that give ^{1}H NMR signals **shifted upfield** from TMS: negative δ values.

Upfield and downfield are directions relative to where you point your finger on the NMR chart. Signals to the *right* of where you are are **upfield.** Signals to the *left* of where you are are **downfield.**

FIGURE 33.3 NMR of ethylbenzene.

The Chemical Shift

You can see that not all peaks fall in the same place, so the protons must be in different surroundings. There is one signal at δ 1.23, one at δ 2.75, and another at δ 7.34. You usually take these values from the *center* of a **split signal** (that's coming up). See that the TMS is really zero before you report the chemical shift. If it is not at zero, you'll have to add or subtract one correction from all of the values. This is the same as using a polystyrene calibration peak to get an accurate fix on IR peaks.

You'll need a correlation table or a correlation chart (Fig. 33.4) to help interpret your spectrum. The —CH_3 group is about in the right place (δ 1.23 in Figure 33.3). The δ 7.34 signal is from the aromatic ring, and, sure enough, that's where signals from aromatic rings fall. The δ 2.75 signal from the —CH_2— is a bit trickier to interpret. The chart shows a —CH_3 on a benzene ring in this area. Don't be literal and argue that *you don't have a* —CH_3, *you have a* —CH_2—CH_3. All right, they're different. But the —CH_2— group *is on a benzene ring and attached to a* —CH_3. That's why those —CH_2— protons are farther downfield; that's why you don't classify them with ordinary R—CH_2—R protons. Use some sense and judgment.

I've blocked out related groups on the correlation table in Fig. 33.4. Look at the set from δ 3.1 to δ 4.0. They're the area where protons on carbons attached to halogens fall in. Read that again. It's protons on *carbons* attached to halogens. The more electronegative the halogen on the carbon, the farther *downfield* the *chemical shift of those protons*. The electronegative halogen draws electrons from the carbon and thus from around the protons on the carbon. These protons now don't have as many electrons surrounding them. They are *not as shielded* from the big bad magnetic field as they might be. They are **deshielded,** so their signal falls *downfield*.

The hydrogen-bonded protons wander all over the lot. Where you find them and how sharp their signals are depend at least on the solvent, the concentration, and the temperature.

Some Anisotropy

So what about aromatic protons (δ 6.0–9.5), aldehyde protons (δ 9.5–9.6), or even protons on double, nay, triple bonds (δ 2.5–3.1)? All of these protons are attached to carbons with bonds, double or triple bonds, or aromatic systems. The electrons in these π bonds *generate their own little local magnetic field*. This local field is not *spherically symmetric;* it can shield or deshield protons depending on where the protons are—it's **anisotropic.** In Figure 33.5, the **shielding regions** have pluses on them, and **deshielding regions** have minuses.

This is one of the quirks in the numbering system. Physically and psychologically, a minus means *less* (*less shielding*), and DOWNfield is farther left on the paper; yet the value of δ goes UP. Another system uses the Greek tau (τ)—that's $10.0 - \delta$. So δ 0.00 (ppm) is 10.0 τ. Don't confuse these two systems. And don't ever confuse deshielding (or shielding) with the *proper direction of the chemical shift*.

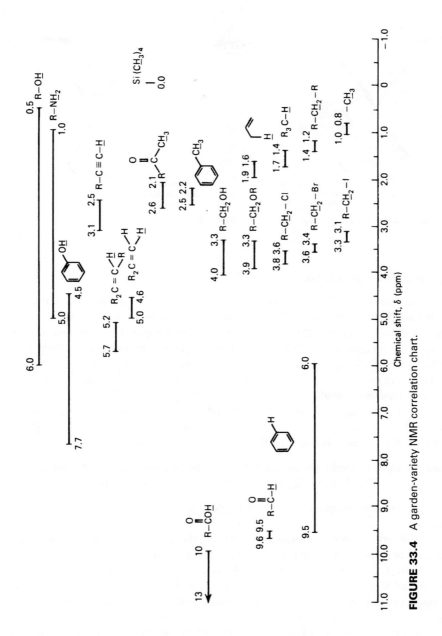

FIGURE 33.4 A garden-variety NMR correlation chart.

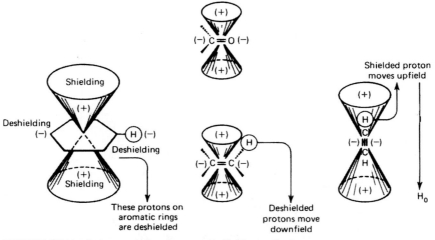

FIGURE 33.5 Anisotropic local magnetic fields on display.

Spin–Spin Splitting

Back to ethylbenzene (Fig. 33.3). You'll find that the —CH_2— and the —CH_3 protons are not single lines; they are *split*. **Spin–spin splitting** is the fancy name. Protons have a spin of $+\frac{1}{2}$ or $-\frac{1}{2}$. If I'm sitting on the methyl group, I can see two protons on the adjacent carbon (—CH_2—). (**Adjacent carbon,** remember that.) They spin, so they produce a magnetic field. Which way do they spin? That's the crucial point. Both can spin one way, *plus*. Both can spin one way, *minus*. Or, each can go a different way: one plus, one minus.

Over at the methyl group (**adjacent carbon,** right?), you can feel these fields. They add a little, they subtract a little, and they cancel a little. So your methyl group splits into three peaks! It's split by the *two* protons *on the adjacent carbon*.

Don't confuse this with the fact that there are three protons on the methyl group! That has nothing to DO with it! It is mere coincidence.

The methyl group shows up as a **triplet** because it is split *by two* protons on the *adjacent carbon*.

Now what about the intensities? Why is the middle peak larger? Get out a marker and draw an A on one proton and a B on the other in the —CH_2— group (Fig. 33.6). OK. There's only one way for A and B to spin in the same direction: *Both A and B are plus, or both A and B are minus.* But there are *two ways* for them to spin opposite each other: *A plus with B minus; B plus with A minus.* This condition happens *two times.* Both A and B plus happen only *one time.* Both A and B minus happen only *one time.* So what? So the **ratio of the intensities is 1 : 2 : 1.** Ha! You got it—a **triplet.** Do this whole business sitting on the —CH_2— group. You get a

FIGURE 33.6 Spin alignments for the ethyl group.

quartet—*four lines—because the —CH₂— protons are adjacent to a methyl group.* They are split *by THREE to give FOUR lines* (Fig. 33.6).

No, this is not all. You can tell that the —CH₂— protons and the —CH₃ protons split each other by their **coupling constant,** the distance between the split peaks of a single group. Coupling constants are called *J* values and are usually given in hertz (Hz). You can read them right from the chart, which has a grid calibrated in hertz. If you find protons at different chemical shifts and their coupling constants are the same, they're splitting and coupling with each other.

Integration

Have you wondered about those funny curves drawn over the NMR peaks? They're **electronic integrations,** and they can tell you how many protons there are at each chemical shift. Measure the distances between the horizontal lines just before and just after each group. With a cheap plastic ruler, I get 52 mm for the benzene ring protons, 21 mm for the —CH₂— protons, and about 30 mm for the —CH₃ protons (see Figure 33.3). Now you divide all the values by the smallest one. Well, 21 mm is the smallest, and without a calculator I get 2.47 : 1 : 1.43. Not even close. And how do you get that 0.47 or 0.43 proton? Try for the simplest *whole number ratio.* Multiply everything by 2, and you'll have 4.94 : 2 : 2.86. This is very close to 5 : 2 : 3, the actual number of protons in ethylbenzene. Use other whole numbers; the results are not so good, and you can't justify the splitting pattern—3 *split* BY 2 and 2 *split* BY 3—with other ratios. Don't use each piece of information in a vacuum.

A Final Note

There are a lot of other things in a typical NMR. There are **spinning sidebands,** small duplicates of stronger peaks evenly spaced from the parent peak. They fall at *multiples of the spin rate,* which in Figure 33.3 is about 30 Hz. Spin the sample tube faster, and these sidebands move farther away; slow the tube and they must get closer.

Signals that split each other tend to lean toward each other (see Figure 33.3 one last time). It's really noticeable in the triplet and even distorts the intensity ratio in the quartet somewhat. Ask your instructor, or see another textbook if you have questions.

EXERCISES

1. Some folks give the gyromagnetic ratio for the proton as $26.75 \times 10^7 \, T^{-1}s^{-1}$ rather than $4.26 \times 10^7 \, Hz/T$, yet both wind up with the same energies for the same systems. How is this done?

2. What are the differences—if any—between NMR sample prep for a CW (continuous wave) instrument and that for an FT instrument?

3. Should you run a sample in chloroform using the chloroform peak as your reference?, What problems might you encounter?

4. Your NMR spinner is a cylinder with about one-quarter of it having a larger diameter than the rest. When you put this on the NMR tube, which end is nearest the cap? If you knew that the tube and spinner would be stuck in the NMR probe if you put it on the wrong way, who would you ask for the proper orientation: another student or your instructor? Comment upon the efficacy of each approach.

5. High-frequency (300 MHz and up) NMR instruments have been said to be a magnet, surrounded by liquid helium, surrounded by liquid nitrogen, surrounded by a postdoc. What's a postdoc? Compare the care of a supercon NMR with that of cows on a farm in the context of being able to take a vacation.

THEORY OF DISTILLATION

Distillation is one of the more important operations you will perform in the organic chemistry laboratory. It is important that you understand some of the physical principles working during a distillation. Different systems require different treatments; the explanations that follow parallel the classifications of distillations given earlier in Chapters 19 and 20.

CLASS 1: SIMPLE DISTILLATION

In a simple distillation, you recall, you separate liquids boiling BELOW 150°C at 1 atm from

1. Nonvolatile impurities.
2. Another liquid boiling at least 25°C higher than the first. The liquids should dissolve in each other. If they do not, then you should treat the system like a steam distillation, and if you're going to steam distill, be sure to look at the later discussion for "Class 4: Steam Distillations." The boiling points should differ by 25°C because you may assume that the higher-boiling component doesn't do anything but sit there during the distillation. Otherwise, you may have a two-component system, and you need to look at the discussion for "Class 3: Fractional Distillation" as well as the discussion here.

That said, let's go on with our discussion of the distillation of one-component systems.

Suppose you prepared isobutyl alcohol by some means, and at the end of the reaction you wound up with a nice yellow-brown liquid. Checking any handbook, you find that isobutyl alcohol is a colorless liquid that boils at 108.1°C at 760 torr. On an STP day, the atmospheric pressure (P) is 760 torr and the boiling point is the **normal boiling point.** At that point, the single point where the vapor pressure of the liquid is the same as that of the atmosphere, the liquid boils.

Fearing little, you set up a Class 1 simple distillation and begin to heat the mix. If you kept track of the temperature of the *liquid* (and you don't; the thermometer bulb is up *above* the flask) and its *vapor pressure,* you'd get the temperature–vapor pressure data in columns 1 and 2 of Table 34.1.

At 82.3°C the vapor pressure of the liquid is 290.43 torr—much *lower* than atmospheric pressure. The liquid doesn't boil. Heat it to, say, 103.4°C, and the vapor pressure is 644.16 torr. Close to atmospheric pressure, but no prize. Finally, at 108.1°C, we have 760.04 torr. *The vapor pressure of the liquid equals that of the atmosphere, and the liquid boils.*

Now, do you see an entry in the table for brown gunk? Of course not. The brown gunk *must* have a very low (or no) vapor pressure at any temperature you might hit during your distillation. Without a vapor pressure, there can be no vapor. No vapor, and there's nothing to condense. Nothing to condense, and there's no distillation. So the isobutyl alcohol comes over clean and pure, and the brown gunk stays behind.

TABLE 34.1   Temperature–Vapor Pressure Data for Isobutyl and Isopropyl Alcohol

Temperature (°C)	Vapor pressure (torr)	
	Isobutyl alcohol	Isopropyl alcohol
82.3	290.43	**760.00**
83.2	301.05	786.31
85.4	328.41	853.90
86.9	348.27	902.76
88.7	373.46	964.52
90.0	406.34	1044.83
95.8	488.59	1244.30
99.9	567.96	1435.11
103.4	644.16	1616.99
106.2	711.23	1776.15
108.1	**760.04**	1891.49

If you plot the temperature and vapor pressure data given in Table 34.1, you reconstruct the **liquid–vapor equilibrium line** in the **phase diagram** of that liquid (Fig. 32.1). The equation of this line, and you might remember this from your freshman chemistry course, is the *Clausius–Clapeyron* equation.

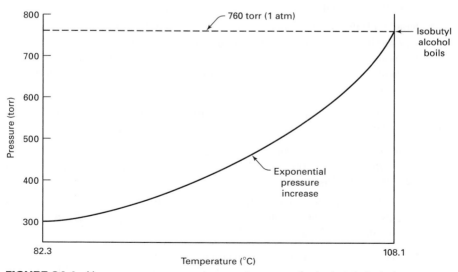

FIGURE 34.1   Vapor pressure versus temperature curve for isobutyl alcohol.

Clausius and Clapeyron

If you want to know how the vapor pressure of a substance is going to vary with temperature, you can use the *Clausius–Clapeyron* equation:

$$p = p°\exp\left[\frac{-\Delta H}{R}\left(\frac{1}{T} - \frac{1}{T°}\right)\right]$$

where

$p°$ is a known vapor pressure

$T°$ is the known temperature (in K, *not* °C)

ΔH is the heat of vaporization of the liquid

R is the universal gas constant (8.314 J/mol–K)

T is the temperature for which you want the vapor pressure

p is the vapor pressure that you calculate for the temperature T that you want

The values for $p°$ and $T°$ are usually taken from the normal boiling point. I've put this formula to two uses:

1. From the normal boiling point of isobutyl alcohol ($T° = 108.1$°C or 381.2 K, $p° = 760$ torr) and one other vapor pressure measurement that I found while doing research for this section ($T = 100$°C or 373 K, $p = 570$ torr), I've gotten the **heat of vaporization (ΔH).** This is the heat needed to vaporize a mole of pure isobutyl alcohol itself. Using the values for $T°$, $p°$, T, and p in the Clausius–Clapeyron equation, I found that the ΔH for isobutyl alcohol is 10,039.70 cal/mol.

2. Now, with the ΔH for isobutyl alcohol, I've calculated the field of pressures you see in Table 34.1 from temperatures of 82.3° to 108.1°C. *That's why the last pressure at 108.1 is 760.04, and not 760.00. The 760.04 value has been back-calculated using the ΔH that we calculated from the top pressure–temperature points in the first place.*

I've also listed the vapor pressure data for isopropyl alcohol in column 3 of Table 34.1. Again, two steps were required to generate the data:

1. Two known vapor pressure–temperature points ($T° = 82.3$°C or 355.4 K, $p° = 760$ torr; $T = 100$°C or 373 K, $p = 1440$ torr) were used to calculate the ΔH: 9515.73 cal/mol.

2. That ΔH for isopropyl alcohol and temperatures from 82.3°C to 108.1°C were then used to calculate a field of vapor pressures for isopropyl alcohol. These are in column 3 of Table 34.1.

I've done these calculations for several good reasons:

1. To show you how to use the Clausius–Clapeyron equation and to show you how well the equation fits over small temperature ranges. The calculated boiling-point pressure for isobutyl alcohol (760.04 torr) is not very different from the normal boiling-point pressure of 760.00 torr (they're 0.005% different, if you must know).

2. To show you that compounds with *higher* Δ*H values have lower vapor pressures*. This means that it takes more energy to vaporize them.

3. To show you that you can calculate vapor pressures that are *above* the boiling point of the liquid. They have a slightly different meaning, however. There is no liquid isopropyl alcohol at 100°C and 760 torr. The vapor pressure at 100°C is 1440 torr, almost twice the atmospheric pressure. But if we artificially increased the pressure over a sample of isopropyl alcohol (pumped up the flask with compressed air?) to 1440 torr and then heated the flask, the alcohol would no longer boil at 82.3°C. You'd have to go as high as—did someone say 100°C?—before the vapor pressure of the liquid matched the now pumped-up atmospheric pressure, and the liquid would boil.

4. To show you the theory of the next topic, Class 3: Fractional Distillation.

CLASS 3: FRACTIONAL DISTILLATION

In a fractional distillation, you remember, you are usually separating liquid mixtures, soluble in one another, that boil at less than 25°C from each other at a pressure of 1 atm.

Now, I didn't discuss *both* isopropyl and isobutyl alcohol in the last section for my health. Suppose you're given a mixture of these two to separate. They *are* miscible in each other, and their boiling points are just about 25°C apart. A textbook case, eh?

A Hint from Dalton

So you set up a fractional distillation and begin to heat the liquid mixture. After a bit, it boils. And what does that mean? The vapor pressure of the *solution* (not just one component) is now equal to the atmospheric pressure, 760 torr. (We're very lucky that textbook-land has so many STP days.) Each component exerts its own vapor pressure, and when the *total* pressure reaches 760 torr, the solution boils.

$$P_{total} = P_A + P_B$$

This is **Dalton's law of partial pressures.** The total pressure of the gang is equal to the sum of their individual efforts. Here, A could be the isopropyl alcohol and B the isobutyl alcohol (it doesn't matter), but P_{total} must be the *atmospheric pressure, P_{atm}.* So a special version of Dalton's law of partial pressures for use in fractional distillation will be

$$P_{atm} = P_A + P_B$$

Dalton and Raoult

If that's all there were to it, we'd be talking about Class 4: Steam Distillations and the like, where the components aren't soluble, and we could quit. Here, *the two are soluble in each other*. The individual vapor pressures of each component (P_A, P_B) depend not only on the temperature, but also on their **mole fraction.**

It makes sense, really. Molecules are the beasties escaping from solution during boiling, and, well, if the two liquids dissolve in each other perfectly, the more molecules (moles) of one component you have, the more the solution behaves like that one component, until it gets to be the same as a one-component liquid. This is Raoult's law,

$$P_A = X_A P_A^\circ$$

where

> P_A is the vapor pressure of A from the mixture
>
> X_A is the mole fraction of liquid A
>
> P_A° is the vapor pressure of the pure liquid A

If we change the As to Bs, can you still follow me? It's the same thing, only now with liquid B. If we combine the special case of Dalton's law with Raoult's law, we get

$$P_{atm} = X_A P_A^\circ + X_B P_B^\circ$$

Look at this. If there is NO B, then the fraction of A is 1, and the pure liquid A boils when its vapor pressure equals the atmospheric pressure. Didn't I say that? Similarly, for B without A, the mole fraction of B is 1, and it too boils when its vapor pressure equals the atmospheric pressure.

At this point, you're usually given the temperature-versus-mole-fraction diagram for two miscible liquids (Fig. 34.2), and you're told it's a consequence of Raoult's law. Well, yes. But not directly. Raoult's law is a relationship of *pressure,* not temperature, versus mole fraction, and this law is pretty much a straight line. You don't need all your orbitals filled to see that you've been presented with a *temperature-versus-mole-fraction* diagram, there are *two lines* (not one), and neither of them is very straight.

A Little Algebra

I want to convert the combined laws of Dalton and Raoult so that I can show the variation in mole fraction explicitly. First, you'll agree that the mole fractions of A and B *must* add to 1 (or they wouldn't be fractions), so

$$X_A + X_B = 1$$

Now, back with Dalton and Raoult,

$$P_{atm} = X_A P_A^\circ + X_B P_B^\circ$$

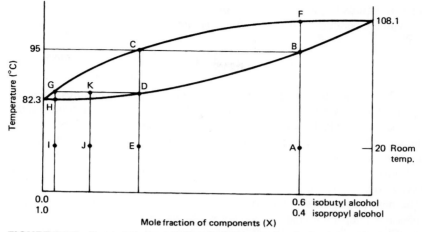

FIGURE 34.2   Temperature–mole fraction diagram for the isobutyl–isopropyl alcohol system.

and seeing that $X_B = 1 - X_A$, I substitute to get

$$P_{atm} = X_A P_A^\circ + (1 - X_A)P_B^\circ$$

Expand this expression with a multiplication:

$$P_{atm} = X_A P_A^\circ + P_B^\circ - X_A P_B^\circ$$

Collect the terms with the mole fraction in them:

$$P_{atm} = X_A P_A^\circ - X_A P_B^\circ + P_B^\circ$$

and factor the mole fraction out to give

$$P_{atm} = X_A(P_A^\circ - P_B^\circ) + P_B^\circ$$

To isolate the mole fraction X_A, subtract P_B° from both sides:

$$P_{atm} - P_B^\circ = X_A(P_A^\circ - P_B^\circ)$$

and divide by $(P_A^\circ - P_B^\circ)$ to get

$$X_A = \frac{P_{atm} - P_B^\circ}{(P_A^\circ - P_B^\circ)}$$

Clausius and Clapeyron Meet Dalton and Raoult

We still have a formula that relates mole fraction to pressure. Note, however, that with the exception of the atmospheric pressure (P_{atm}), *all* the other pressures are those of pure liquids (P_A° and P_B°). Now, how does the vapor pressure of a pure liquid vary with temperature? Smite your forehead and say, "The *Clausius–Clapeyron* equation":

$$p = p^\circ \exp\left[\frac{-\Delta H}{R}\left(\frac{1}{T} - \frac{1}{T^\circ}\right)\right]$$

The P°s of Dalton–Raoult are vapor pressures taken at fixed temperatures. They are the ps in the Clausius–Clapeyron equation found with a variation in temperature. You don't believe me? Pick a vapor pressure and temperature pair from Table 34.1 for either liquid, and let these be p° and T° (and don't forget to use K, not °C). Now what happens when the "unknown" temperature (T) is the same as T°? The ($1/T - 1/T^\circ$) becomes zero, the entire exponent becomes zero, p° is multiplied by 1 (anything to the power zero is 1, eh?), and so $p = p^\circ$.

The next part is messy, but somebody's got to do it. I'm going to use the vapor pressure–temperature data for the normal boiling points of both liquids in the Clausius–Clapeyron equation. Why? They're convenient, *known* vapor pressure–temperature points. When I do this, however, I exercise my right to use different superscripts to *impress on you that these points are the normal boiling points*. So, for liquid A, we have P_A° and T_A°; if A is isobutyl alcohol, $P_A^\circ = 760$ torr and $T_A^\circ = 101.8$°C (374.9 K). For liquid B, we have P_B° and T_B°; if B is isopropyl alcohol, $P_B^\circ = 760$ torr and $T_B^\circ = 82.3$°C (355.5 K).

Using the new symbols and substituting the Clausius–Clapeyron equation for every P°, you get

$$X_A = \frac{P_{atm} - P_B^\circ \exp[(-\Delta H_B/R)(1/T - 1/T_B^\circ)]}{P_A^\circ \exp[(\Delta H_A/R)(1/T - 1/T_A^\circ)] - P_B^\circ \exp[(-\Delta H_B/R)(1/T - 1/T_B^\circ)]}$$

Whew!

The only variables in this beast are the mole fraction (X_A) and the temperature (T). Every other symbol is a constant. Finally, we have the equation for the bottom line of the temperature–mole fraction diagram (Fig. 34.2), something that has eluded us for years.

Dalton Again

What about the upper curve in Figure 34.2? Glad you asked (sigh). The composition in the vapor is also related to Dalton's law of partial pressures. For an ideal gas, $PV = nRT$ (and you thought you'd never see that again!), and for the vapors above the liquid,

$$P_A = n_A \frac{RT}{V}$$

and

$$P_B = n_B \frac{RT}{V}$$

Yet we know that the total pressure of Dalton's gang is the sum of their individual efforts:

$$P_{total} = P_A + P_B$$

So,

$$P_{total} = n_{total}\left(\frac{RT}{V}\right)$$

Now watch as I divide the pressure of A by the total pressure:

$$P_A/P_{total} = n_A/n_{total}\left(\frac{RT/V}{RT/V}\right)$$

Well, the RT/V terms cancel, giving

$$P_A/P_{total} = n_A/n_{total}$$

The ratio of the number of moles of A to the *total* number of moles is the **mole fraction of component A in the vapor:**

$$X_A^{vapor} = P_A/P_{total}$$

P_{total} for the ordinary distillation is the atmospheric pressure, 760 torr. P_A is the vapor pressure of A, and, again by Raoult's law, $P_A = X_A P_A^\circ$. Putting the two together, we get

$$X_A^{vapor} = X_A^{liquid}\, P_A^\circ/760$$

We can make the same kind of substitution of Clausius–Clapeyron here, and we get a similarly curved function for the upper line in the temperature–mole fraction diagram.

To show you that all this really does work, I've listed the experimental composition data for the isopropyl/isobutyl alcohol system from Landolt–Bornstein (Landolt–Bornstein is to physical chemistry what Beilstein is to organic), along with my calculated data (Table 34.2). (That explains my choice of temperatures for Table 34.1.) I've also given the absolute and percent differences between the experimental data and what I've calculated. These differences are on the order of 1% or less, very good agreement indeed.

So now, for any two liquids, if you have their normal boiling points and vapor pressures at any other temperature, you can generate the temperature–mole fraction diagram.

What Does It All Mean?

Getting back to the temperature–mole fraction diagram (Fig. 34.2), suppose that you start with a mixture such that the mole fractions are as follows: isobutyl alcohol, 0.60; isopropyl alcohol, 0.40. On the diagram, that composition is point A at a room temperature of 20°C. Now you heat the mixture, and you travel upward from point A to point B. The liquid has the same composition; it's just hotter.

At 95°C, point B, the mixture boils. Vapor, with the composition at point C, comes flying out of the liquid (the horizontal line tying the composition of the vapor

TABLE 34.2 Experimental and Calculated Data[a]

Temperature Versus Mole Fraction Calculations

Compound	Normal bp (°C, Moore)	Vapor pressure at 100°C (Moore)	Calculated ΔH (vap)
2-Propanol	82.3	1,440 torr	9,515.73 cal/mol
2-Methyl-1-propanol	108.2	570 torr	10,039.70 cal/mol

Comparison of Data from Moore and Calculated Data ($T = 100$°C)

Parameter	Compound	Moore	Calc.
Vapor pressure	2-Propanol	1,440	1,440.05
(torr)	2-Methyl-1-propanol	570	570.02
Mole fraction	2-Propanol	0.219	0.2184
(in liquid)	2-Methyl-2-propanol	0.781	0.7861
Mole fraction	2-Propanol	0.415	0.4137
(in vapor)	2-Methyl-2-propanol	0.585	0.5863

Mole Fraction of 2-Propanol Liquid Data

#	T (°C)	X Calc.	X Lit.	Diff.	% Diff.
1	83.2	0.9458	0.9485	−0.0027	−0.286
2	85.4	0.8213	0.8275	−0.0062	−0.748
3	86.9	0.7425	0.7450	−0.0025	−0.331
4	88.7	0.6540	0.6380	0.0160	2.505
5	90.9	0.5539	0.5455	0.0084	1.540
6	95.8	0.3591	0.3455	0.0136	3.949
7	99.9	0.2215	0.2185	0.0030	1.357
8	103.4	0.1191	0.1155	0.0036	3.096
9	106.2	0.0458	0.0465	−0.0007	−1.507

Mole Fraction of 2-Propanol Vapor Data

#	T (°C)	X Calc.	X Lit.	Diff.	% Diff.
1	83.2	0.9785	0.9805	−0.0020	−0.201
2	85.4	0.9228	0.9295	−0.0067	−0.723
3	86.9	0.8820	0.8875	−0.0055	−0.618
4	88.7	0.8300	0.8245	0.0055	0.663
5	90.9	0.7615	0.7580	0.0035	0.460
6	95.8	0.5880	0.5845	0.0035	0.600
7	99.9	0.4182	0.4270	−0.0088	−2.063
8	103.4	0.2533	0.2510	0.0023	0.936
9	106.2	0.1070	0.1120	−0.0050	−4.433

[a]Literature data from Moore, 3rd ed.

to the composition of the liquid is the **liquid–vapor tie-line**), and this vapor condenses (point *C* to point *D*), say, part of the way up your distilling column.

Look at the composition of this *new liquid* (point *E*). It is *richer in the lower-boiling component*. The step cycle *B–C–D* represents one distillation.

If you heat this new liquid that's richer in isopropyl alcohol (point *D*), you get vapor (composition at point *G* along a horizontal tie-line) that condenses to liquid *H*. So step cycle *D–G–H* is another distillation. The two steps represent two distillations.

Much of this work was carried out using a special distilling column called a **bubble-plate column** (Fig. 34.3). Each *plate* really does act like a distilling flask with a very efficient column, and one distillation is really carried out on one physical plate. To calculate the number of plates (separation steps or distillations) for a bubble-plate column, you just count them!

Unfortunately, the fractionating column you usually get is not a bubble-plate type. You have an open tube that you fill with column packing (see "Class 3: Fractional Distillation" in Chapter 19) and *no plates*. The distillations up this type of column are not discrete, and the question of where one plate begins and another ends is meaningless. Yet, if you use this type of column, you do get a better separation than if you used no column at all. It's *as if* you had a column with some bubble plates. And if your distilling column separates a mixture as well as a bubble-plate column with two real plates, you must have a column with two **theoretical plates.**

You can calculate the number of theoretical plates in your column if you distill a two-component liquid mixture of known composition (isobutyl and isopropyl alcohol, perhaps?) and collect a few drops of the liquid condensed from the vapor at the top of the column. You need to determine the composition of that condensed vapor, usually from a calibration curve of known compositions versus their refractive indices (see Chapter 29, "Refractometry"), and you must have the temperature–mole fraction diagram (Fig. 34.2).

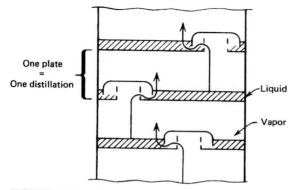

One plate
=
One distillation

Liquid

Vapor

FIGURE 34.3   A bubble-plate fractionating column.

Suppose you fractionated that liquid of composition A, collected a few drops of the condensed vapor at the top of the column, analyzed it by taking its refractive index, and found that this liquid had a composition corresponding to point *J* on our diagram. You would follow the same path as before (*B–C–D*, one distillation; *D–G–H*, another distillation) and find that the composition falls a bit short of the full cycle for distillation number 2.

Well, all you can do is estimate that it falls at, say, a little more than halfway along this second tie-line (point *K*), eh? OK then. This column has been officially declared to have 1.6 theoretical plates. Can you have tenths of plates? Not with a bubble-plate column, but certainly with any column that does not have discrete separation stages.

Now you have a column with 1.6 theoretical plates. "Is that good?" you ask. "Relative to what?" I say. If that column is 6 feet high, that's terrible. The height equivalent to a theoretical plate (HETP) is 3.7 ft/plate. Suppose that another column also had 1.6 theoretical plates, but was only 6 in. (0.5 ft) high. The HETP for this column is 3.7 in./plate, and *if it were 6 ft high,* it would have 19 plates. The *smaller* the HETP, the more efficient the column is. There are more plates for the same length.

One last thing. There's a point *F* on the temperature–mole fraction diagram in Fig. 34.2 that I haven't bothered about. *F* is the grade you'll get when you extend the *A–B* line up to cut into the upper curve and you then try to do anything with this point. I've found an amazing tendency for some people to extend that line to point *F*. Why? Raise the temperature from *A* to *B*, and the sample boils. *When the sample boils, the temperature stops going up.* Heat going into the distillation is being used to vaporize the liquid (heat of vaporization, eh?), and all you get is a vapor, enriched in the lower-boiling component, with the composition found at the end of a *horizontal* tie-line.

Reality Intrudes I: Changing Composition

To get the number of theoretical plates, we fractionally distilled a known mixture and took off a small amount for analysis, so as not to disturb things very much. You, however, have to fractionally distill a mixture and hand in a good amount of product, and do it within the time limits of the laboratory.

So when you fractionally distill a liquid, you *continuously* remove the lower-boiling fraction from the *top* of the column. And where did that liquid come from? The boiling fluid at the *bottom* of the column. Now, if the distillate is richer in the lower-boiling component, what happened to the composition of the boiling liquid? I'd better hear you say that the boiling liquid gets richer in the *higher-boiling* component (Fig. 34.4).

So as you fractionally distill, not only does your boiling liquid get richer in the higher component, but so also does your distillate, your condensed vapor. Don't worry too much about this effect. It happens as long as you have to collect a product for evaluation. Let your thermometer be your guide, and keep the temperature spread less than 2°C for any product you collect.

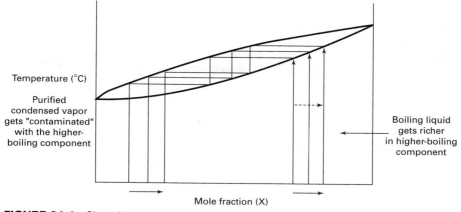

FIGURE 34.4   Changing composition as the distillation goes on.

Reality Intrudes II: Nonequilibrium Conditions

Not only were we forced to remove a small amount of liquid to accurately determine the efficiency of our column, we also had to do it *very slowly*. This allowed the distillation to remain *at equilibrium*. The **throughput,** the rate at which we took material out of the column, was *very low*. Of all the molecules of vapor that condensed at the top of the column, *most fell back down the column; few were removed*—a very high **reflux ratio.** With an infinite reflux ratio (no throughput), the condensed vapor at the top of the column is as rich in the lower-boiling component as it's ever likely to get in your setup. As you remove this condensed vapor, the equilibrium is upset, as more molecules rush in to take the place of the missing. The faster you distill, the less time there is for equilibrium to be reestablished, and the less time there is for the more volatile components to sort themselves out and move to the top of the column. So you begin to remove higher-boiling fractions as well, and you cannot get as clean a separation. In the limit, you could remove vapor so quickly that you shouldn't have even bothered using a column.

Reality Intrudes III: Azeotropes

Occasionally, you'll run across liquid mixtures that cannot be separated by fractional distillation. That's because the composition of the vapor coming off the liquid is the *same* as that of the liquid itself. You have an **azeotrope,** a liquid mixture with a constant boiling point.

Go back to the temperature–mole fraction diagram for the isopropyl alcohol–isobutyl alcohol system (Fig. 34.2). The composition of the vapor is *always different from* that of the liquid, and we can separate the two compounds. If the composition of the vapor is the same as that of the liquid, that separation is hopeless. Since we've used the notions of an ideal gas in deriving our equations for the liquid and vapor compositions (Clausius–Clapeyron, Dalton, and Raoult), this azeotropic behavior is said to result from *deviation from ideality*, specifically *deviations from Raoult's law*. Although you might invoke certain interactive forces in explaining nonideal behavior, you cannot predict azeotrope formation *a priori*. Very similar materials form azeotropes (ethanol–water). Very different materials form azeotropes (toluene–water). And they can be either **minimum-boiling azeotropes** or **maximum-boiling azeotropes.**

Minimum-Boiling Azeotropes   The ethanol–water azeotrope (95% ethanol–5% water) is an example of a minimum-boiling azeotrope. Its boiling point is lower than that of the components (Fig. 34.5). If you've ever fermented anything and distilled the results in the hopes of obtaining 200-proof (100%) white lightning, you'd have to content yourself with getting the azeotropic 190-proof mixture instead. Fermentation usually stops when the yeast die in their own 15% ethanol solution. At room temperature, this is point *A* on our phase diagram. When you heat the solution, you move from point *A* to point *B*, and (urges to go to point *F* notwithstanding) you cycle through distillation cycles *B–C–D* and *D–E–G*. And, well, guess what comes off the liquid? Yep, the azeotrope. As the azeotrope comes over, the composition of the boiling liquid moves to the right (it gets richer in water), and finally there isn't enough ethanol to support the azeotropic composition. At that point, you're just distilling water. The process is mirrored if you start with a liquid that is 95% ethanol and water. The azeotrope comes off first.

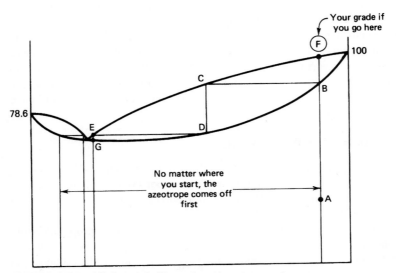

FIGURE 34.5   Minimum-boiling ethanol–water azeotrope.

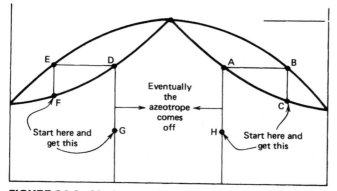

FIGURE 34.6 Maximum-boiling chloroform–acetone azeotrope.

Maximum-Boiling Azeotropes The chloroform–acetone azeotrope (52% chloroform–48% acetone) is an example of the much rarer maximum-boiling azeotrope. Its boiling point is higher than that of the components (Fig. 34.6). At compositions off the azeotrope, you do distillations *A–B–C* or *D–E–F* (and so on) until the boiling liquid composition reaches the azeotropic composition. Then that's all that comes over. So, initially, one of the components comes off first.

Azeotropes on Purpose You might think the formation of azeotropes is an unalloyed nuisance, but they can be useful. Toluene and water form a minimum-boiling azeotrope (20.2% water; 85°C). If for some reason you need *very dry* toluene, all you have to do is distill some of it. The water–toluene azeotrope comes off first, and, well, there goes all the water. It is removed by **azeotropic distillation.** The technique can also be used in certain reactions, including the preparation of amides from very-high-boiling acids and amines. (They can even be solids.) You dissolve the reagents in toluene, set up a reflux condenser fitted with a **Dean–Stark trap** (Fig. 34.7), and let the mixture reflux. As the amide forms, water is released, and the water is constantly removed by azeotropic distillation with the toluene. The azeotrope cools, condenses, and collects in the Dean–Stark trap. At room temperature, the azeotrope is said to "break," and the water forms a layer at the bottom of the trap. Measure the amount of water, and you have an idea of the extent of the reaction.

Absolute (100%) ethanol is often made by adding benzene to the ethanol–water **binary azeotrope** (two components) to make a **ternary azeotrope** (three components). This ternary alcohol–water–benzene (18.5 : 7.4 : 74.1) azeotrope comes over until all the water is gone and is followed by a benzene–ethanol mixture. Finally, absolute ethanol gets its chance to appear, marred only slightly by traces of benzene.

Other Deviations

The furfural–cyclohexane phase diagram (Fig. 34.8) shows that you can have mixtures that exhibit nonideal behavior without having to form an azeotrope. In sum,

Azeotrope
condenses
here

falls
down

and
"breaks"
here

Water
collects

Dried
organic
compound
goes
back
to flask

Azeotrope
goes
up

Wet liquid
boils here

FIGURE 34.7 Removing water by azeotropic distillation.

without the phase diagram in front of you, you shouldn't take the distillation behavior of any liquid mixture for granted.

CLASS 4: STEAM DISTILLATION

Steam distillation is for separating mixtures of tars and oils, and they must not dissolve much in water. If you think about it a bit, this could be considered a fractional distillation of a binary mixture with an *extreme* deviation from Raoult's law. The

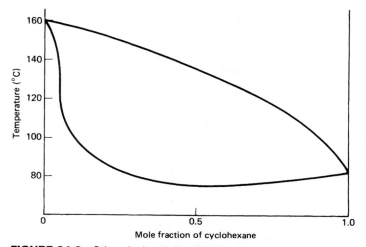

FIGURE 34.8 Other deviant behavior (but no azeotropes) in the furfural–cyclohexane system.

water and the organic oils want nothing to do with each other—so much so that you can consider them unmixed in separate compartments of the distilling flask. As such, they act completely independently of each other. The mole fraction of each component in its own compartments is 1. So Raoult's law becomes

$$P_{\text{total}} = P_A + P_B$$

This is just Dalton's law of partial pressures. P_{total} is P_{atm} for a steam distillation. So the vapor pressure of the organic oil is now less than that of the atmosphere and the water, and the oil codistills at a much lower temperature.

As an example, suppose you were to try to distill quinoline directly. Quinoline has a boiling point of 237°C at 1 atm. Heating organic molecules to these temperatures may often be a way to decompose them. Fortunately, quinoline is insoluble in water, and it does have some vapor pressure at about the boiling point of water (10 torr at 99.6°C). If it had a *much lower* vapor pressure at the boiling point of water (say, 0.1 torr), there couldn't be enough of it vaporizing to make even steam distillation worthwhile.

Well, at 99.6°C, quinoline contributes 10 torr to the total vapor pressure, and water must make up the difference (750 torr) to satisfy Dalton's law of partial pressures to make P_{total} 760 torr at boiling. Using the relationships for the composition of the vapor *over a liquid,* we can calculate the quinoline/water ratio coming over.

If we consider that each is an ideal gas, then

$$PV = nRT \quad \text{(yes, again)}$$

The number of moles (n) of anything is just the weight in grams (g) divided by the *molecular weight* of that substance (MW), and so

$$PV = (g/\text{MW})(RT)$$

Multiplication by MW gives

$$(\text{MW})PV = g(RT)$$

and isolating the mass of the material by dividing by RT gives

$$(\text{MW})PV/RT = g$$

Now for *two* vapors, A and B, I'll construct a ratio where

$$\frac{(\text{MW})_A P_A V/RT = g_A}{(\text{MW})_B P_B V/RT = g_B}$$

With A and B in the same flask, R, T, and V must be the same for each, and we can cancel these terms to get

$$\frac{(\text{MW})_A P_A = g_A}{(\text{MW})_B P_B = g_B}$$

If we plug in values for the molecular weights and vapor pressures of quinoline and water, we get

$$\frac{(129)(10)}{(18)(750)} = \frac{0.0956}{1}$$

So, as an approximation, for every 10 g of distillate we collect, 1 g will be our steam-distilled quinoline.

into ethanol from water, forgetting for the moment that ethanol is *soluble in water* and that you *must have two insoluble liquids* to carry out an extraction. Chagrined, you forget the ethanol and choose diethyl ether.

From the preceding solubility data we can calculate the **distribution,** or **partition coefficient,** for oxalic acid in the water–ether extraction. This coefficient (number) is just the ratio of solubilities of the compound you wish to extract in the two layers. Here,

$$K_p = \frac{\text{solubility of oxalic acid in ether}}{\text{solubility of oxalic acid in water}}$$

which amounts to 16.9/9.5, or 1.779.

Imagine that you have 40 g of oxalic acid in 1000 mL water and you put that in contact with 1000 mL ether. The oxalic acid *distributes itself between the two layers.* How much is left in each layer? Well, if we let x g equal the amount that stays in the water, 1.779x g of the acid has to walk over to the ether. And so

Wt of oxalic acid in ether = (1000 mL)(1.779x g/mL) = 1779x g

Wt of oxalic acid in water = (1000 mL)(x g/mL) = 1000x g

The total weight of the acid is 40 g (now partitioned between two layers) and

$$2779x \text{ g} = 40 \text{ g}$$
$$x = 0.0144$$

and

Wt of oxalic acid in ether = 1779(0.0144) g = 25.6 g

Wt of oxalic acid in water = 1000(0.0144) g = 14.4 g

Now, let's start with the same 40 g of oxalic acid in 1000 mL of water, but this time we will *do three extractions with 300 mL of ether*. The first 300-mL portion hits, and

Wt of oxalic acid in ether = (300 mL)(1.779x g/mL) = 533.7x g

Wt of oxalic acid in water = (1000 mL)(x g/mL) = 1000x g

The total weight of the acid is 40 g (now partitioned between two layers) and

$$1533.7x \text{ g} = 40 \text{ g}$$
$$x = 0.0261$$

so

$$\text{Wt of oxalic acid in ether} = 533.7(0.0261) \text{ g} = 13.9 \text{ g}$$
$$\text{Wt of oxalic acid in water} = 1000(0.0261) \text{ g} = 26.1 \text{ g}$$

That ether layer is removed, and the *second jolt* of 300 mL fresh ether hits, and

$$\text{Wt of oxalic acid in ether} = (300 \text{ mL})(1.779x \text{ g/mL}) = 533.7x \text{ g}$$
$$\text{Wt of oxalic acid in water} = (1000 \text{ mL})(x \text{ g/mL}) = 1000x \text{ g}$$

But here we started with 26.1 g of acid in water (now partitioned between two layers) and

$$1533.7x \text{ g} = 26.1 \text{ g}$$
$$x = 0.0170$$

so

$$\text{Wt of oxalic acid in ether} = 533.7(0.0170) \text{ g} = 9.1 \text{ g}$$
$$\text{Wt of oxalic acid in water} = 1000(0.0170) \text{ g} = 17.0 \text{ g}$$

Again, *that ether layer is removed,* and the *third jolt* of 300 mL fresh ether hits, and

$$\text{Wt of oxalic acid in ether} = (300 \text{ mL})(1.779x \text{ g/mL}) = 533.7x \text{ g}$$
$$\text{Wt of oxalic acid in water} = (1000 \text{ mL})(x \text{ g/mL}) = 1000x \text{ g}$$

But here, we started with 17.0 g of acid in water (now partitioned between two layers) and

$$1533.7x \text{ g} = 17.0 \text{ g}$$
$$x = 0.011$$

so

$$\text{Wt of oxalic acid in ether} = 533.7(0.011) \text{ g} = 5.87 \text{ g}$$
$$\text{Wt of oxalic acid in water} = 1000(0.011) \text{ g} = 11.0 \text{ g}$$

(They don't quite add up to 17.0 g—I've rounded them off a bit.)

Let's consolidate what we have: first, 13.9 g, then 9.1 g, and finally 5.87 g of oxalic acid, for a total of 28.9 g (72.3%) of acid extracted into 900 mL of ether. OK, that's not far from 25.6 g (64%) extracted *once* into 1000 mL of ether. That's because the distribution coefficient is fairly low. But it *is* more. That's because *several small extractions are better than one large one.*

INDEX